Briefings on Existence

SUNY series, Intersections: Philosophy and Critical Theory
Rodolphe Gasché, Editor

Briefings on Existence

A Short Treatise on Transitory Ontology

Alain Badiou

TRANSLATED, EDITED,
AND WITH AN INTRODUCTION BY

Norman Madarasz

State University of New York Press

Originally published in France
under the title *Court Traité d'ontologie transitoire*
Copyright: © 1998, Éditions du Seuil

English translation made by agreement with Éditions du Seuil

Published by
State University of New York Press, Albany

English translation © 2006 State University of New York

Printed in the United States of America

For information, address State University of New York Press
194 Washington Avenue, Suite 305, Albany, NY 12210-2384

Production by Diane Ganeles
Marketing by Susan M. Petrie

Library of Congress Cataloging-in-Publication Data

Badiou, Alain.
    [Court traité d'ontologie transitoire. English]
    Briefings on existence : a short treatise on transitory ontology / Alain
Badiou ; translated, edited and with an introduction by Norman Madarasz.
        p. cm. — (SUNY series, intersections—philosophy and critical theory)
    Includes bibliographical references and index.
    ISBN-13: 978-0-7914-6803-6 (hardcover : alk. paper)
    ISBN-10: 0-7914-6803-8 (hardcover : alk. paper)
    ISBN-13: 978-0-7914-6804-3 (pbk. : alk. paper)
    ISBN-10: 0-7914-6804-6 (pbk. : alk. paper) 1. Ontology. I. Madarasz,
Norman. II. Title. III. Intersections (Albany, NY)

    BD312.B3213   2006
    111—dc22   2005033878
                        10 9 8 7 6 5 4 3 2 1

I call "transitory ontology" the ontology unfolding between the science of Being *qua* Being, that is, the theory of the pure manifold, and the science of appearing, that is, the logic of the consistency of actually presented universes. It is a trajectory of thought. This short book sets a few of its milestones.

—Alain Badiou

CONTENTS

Preface to the English-language Edition / ix

Translator's Introduction:
Alain Badiou: Back to the Mathematical Line / 1

Prologue: God is Dead / 21

1. The Question of Being Today / 33

2. Mathematics is a Thought / 45

3. The Event as Trans-Being / 59

4. Deleuze's Vitalist Ontology / 63

5. Spinoza's Closed Ontology / 73

6. Platonism and Mathematical Ontology / 89

7. The Aristotelian Orientation and Logic / 101

8. Logic, Philosophy, "Linguistic Turn" / 107

9. First Remarks on the Concept of *Topos* / 115

10. First Provisional Theses on Logic / 119

11. The Being of Number / 125

12. Kant's Subtractive Ontology / 133

Contents

13. Group, Category, Subject / 143

14. Being and Appearing / 153

Annex / 169

Notes / 171

Contributors / 183

Index / 185

PREFACE TO THE
ENGLISH-LANGUAGE EDITION

In 1986, upon completing the most important philosophy book I had yet written (*L'Être et l'événement*), I felt myself filled with new resources.[1] Thanks to the main concepts I had created, or re-created (situation, event, Subject, truth procedure, forcing, and so on), I felt myself able to produce a new vision on all kinds of things: poems (Mallarmé or Rimbaud), prosaists (Beckett), po-litical sequences (Saint-Just, Lenin, and Mao), psychoanalytic se-quences (Lacan), and mathematical novelties (on the concept of number). . . . In the few odd years between 1986 and 1992, I mul-tiplied lectures, articles, and sundry public appearances. I was liv-ing in thought like someone who had fallen upon an oil well: an inexhaustible intellectual energy lay at my disposal. This fertile good fortune was recapitulated in the anthology called *Conditions,* published in France in 1992. I had gained very clear insight into what (for me) philosophy was all about. So much so that, as of 1989, I was able to confine my main convictions densely, but in organized fashion, to a short work that made them accessible to a broader public. This was *Manifeste pour la philosophie.* From these convictions I was also able to draw implacable consequences about action and its norms. This book was *L'Éthique, Essai sur la con-science de Mal.* It ended up being quite a best seller in France, and made its way around the planet. So far, it has been translated into sixteen languages.

Yet from the middle of the 1990s, what slowly grew to be-come most evident to me were the difficulties of my undertaking. Happy times were coming to a close. I told myself: "The idea of event is fundamental. But the theory I propose on what the event is the name of is not clear." Or: "The ontological extension of mathematics is certain. But, then, what about logic?" Many other doubts and questions ensued. For example: "How to distinguish an event from an important fact or from a becoming, in Deleuze's sense?" Or: "Do we not have to conceive, if we are materialists, any Subject as having a body? What would that body be, provided it is not the biological body, if not a kind of body of truths?" More difficult still, although more empirical in appearance: "How does one account for the following obvious fact: just as new things can be progressive, so also can they be reactionary. How can it be that something new and novel can also be reactive?" To sharpen my thoughts, I started up a dialogue with thinkers from whom I felt very far in spirit, namely, Nietzsche and Wittgenstein. Then I turned to my great living elder, Gilles Deleuze. We discussed, and that discussion, through various episodes, let to the book *Deleuze. Le clameur de l'être,* published in France in 1997. It, to, quickly became an international best seller. I also returned to a source that was both heterogeneous and very powerful in my view: The apostle Paul. When I organized the results of my reading into book form as *Saint Paul, ou la foundation de l'universalisme,* it became the third best seller.

   All in all, in those years I was leaping from best seller to best seller! Yet the questions I had addressed in *L'Être et l'événement* still lay without a real answer. They kept besetting me as I busied myself at the same time with all kinds of things outside of pure philosophy. From 1993 to 1998, I was constantly doing theater. I wrote a four-play cycle on the Ahmed character and had it staged. That was also the time during which I was writing my novel, *Calme bloc ici-bas.* As anyone will tell you, the prose of novel writing is demanding; it requires very intense concentration. Meanwhile, I had compiled a substantial body of research and information on the most up-to-date mathematics. I had to come to terms with Set Theory's rival theory regarding mathematical foundations: category theory. There were sizeable stakes involved as I had granted Set Theory a key ontological role. I could not exempt myself from proposing a philosophical interpretation of its rival from within my system. Let us add that, as of 1996, I got quite involved in collective action and the thought of that action by organizing a political battle in favor of the *"sans papiers,"* workers who lack authorization to work in France. We strove for them to be fully recognized as laborers living and working in France, and have them obtain the rights entailed by such recognition.

With philosophy, I was going through a middle period. The book you are about to read doubtless bears the difficulties, but also the charm, of this kind of period. Everything in the book is in progress. Nothing is quite concluded. Besides, the title "*transitory ontology*" states this intervallic dimension pretty clearly: on the road with no set destination.

I kept lecturing, penning articles, and exploring every avenue that could be drawn from my philosophical intuitions. I even considered publishing an anthology called *Conditions 2*. But I soon realized that my intellectual situation was very different from what had led to *Conditions* in 1992. At the end of the 1980s, I had taken on certainties. I was experiencing a spirit of conquest. Six or seven years later, I lay amidst questions and doubts, and hard explorations. As it turned out, I was on the road leading me to the book I am finishing. Its title is *Logiques des mondes (World Logics),* subtitled *L'Être et l'événement 2*. Its publication in France is planned for 2006. I believe this *Short Treatise* bears excellent witness to how victory had shifted into doubt and to an obligation, fostered by that doubt, of having to invent other concepts.

Let us note that *Briefings on Existence. Short Treatise on Transitory Ontology* is part of a trilogy, which is already more tentative than the unity of the *Conditions* volume. Published simultaneously with it was the *Petit manuel d'inesthétique,* whose English-language translation has just been published. There was also the *Abrégé de métapolitique,* which should be published in English shortly. *Briefings on Existence* represents the most philosophical moment of that trilogy. I believe it combines the following four principles:

1. Several tightly knit exercises from the history of philosophy (or more accurately, from the history of the thought on Being) devoted to Plato, Aristotle, Spinoza, Kant, Heidegger and Deleuze . . . ;

2. Propositions characteristic of our era, like: the real sense of the statement "God is dead," or the end of the "linguistic turn";

3. A detailed examination of philosophy's relationship to mathematics and formalized logic;

4. A new doctrine of the distinction between "Being" and "appearing" clearly heralding the most important themes of *Logiques des mondes*.

Winding through these elements, I believe, is a red thread. It is the thread of necessity, for which I had somewhat blindly been seeking. I mean the thread that would allow me to keep the trans-temporal ontological intuitions of *L'Être et l'événement* as well as build a complete theory of what appears to us as our world here and now.

My first "big" book asked the question: what about Being, what about truths, and what about the being of truths? In substance, it answered: truths are generic manifolds.

My second "big" book, forthcoming, will ask: what about the world, what about the truth-subjects, and how can these subjects live in a world? You will have to wait until next year to find out the answer it has received.

The *Court traité d'ontologie transitoire*, proposed here to you in English translation as *Briefings on Existence. Short Treatise on Transitory Ontology* by the diligent and friendly care of Norman Madarasz (who also translated and edited *Manifesto for Philosophy*, published by SUNY Press in the "Intersection" series in 1999), bears witness to my efforts of shifting from one book to the next.

This small book is the valorous passage between the big.

Alain Badiou

Notes

1. For all bibliographical references regarding English-language translations of Badiou's work, please see the Translator's Introduction and Notes.

TRANSLATOR'S INTRODUCTION

Alain Badiou:
Back to the Mathematical Line

Norman Madarasz

Nowhere does an author lose control over his text more than in translation. Michel Foucault famously argued that a text buries its author alive by branching out into an uncontrollable array of meaning patterns and interpretations. In translation, however, an author parts with the linear shaping of his oeuvre—all other things being equal.

This has little to do with a sentence, argument, theory, or book falling short of possessing the linguistic tools required to bring a conceptual apparatus into full being in another language and another world. I do not refer here to Heidegger or Ortega y Gasset's infamous declarations on the shortcomings of translating. Instead, it has just about everything to do with the very sequence in which works are translated, which is only compounded when an author writes along several different lines simultaneously.

Alain Badiou has been recognized in France as a leading thinker since the 1960s. Amidst the struggles of the 1970s Badiou's political commitment and continued devotion to egalitarian processes and democratic demands drove him into conflict not merely with the French establishment, but also with the Parisian philosophical milieu. The circumstances around this global conflict resulted in the halt of a short, impressive sequence of publications on mathematical

1

philosophy as the post–1968 climate in France spread toward full-scale social upheaval. His works "La Subversion infinitésimale" and *Le Concept de modèle* were its first achievements.[1]

From Sartre and Godard to Foucault and Bourdieu, science, philosophy, and art were being summoned to converge into providing political transparency to the process of taking the next step in emancipatory becoming, something people used to call "revolution." The urgency of the demands facing off against the quickly moving counteroffensive by right wing political and religious ideologues and the oligarchic business clique shifted interests in France's philosophical scene, steering an entire generation away from academe and toward political activism. In Badiou's case, it also temporarily deprived French thought of a leading mathematical philosopher.

After the mid-1970s the Maoist turn in French politics confronted the catastrophe of Pol Pot's regime. The facile, and often reactionary, decoding of events in southeast Asia, spearheaded by the so-called *nouveaux philosophes'* reading of Solzhenitsyn's *Gulag Archipelago,* vowed to force a connection between France's insistent Left and events in Cambodia. In the same gesture, it conveniently occluded any accusatory mention of the long-term disaster, psychological and environmental, provoked by the illegal secret bombings of the country by the American Air Force, and the extermination occurring shortly thereafter in East Timor, then a territorial holding of the United States' main ally in the Pacific, Indonesia.

Notwithstanding the common front on the French Right, marshaled behind the scenes by Raymond Aron and François Furet, the cataclysm in Cambodia, trial of the Gang of Four in China and inertial centralism of the Soviet system were reshaping the Western European Left's programs and alliances. As Perry Anderson recently wrote, "the uncertainties of the late nineteen-seventies had galvanized into being an 'anti-totalitarian' front that would dominate intellectual life for the next two decades."[2] In turn, the self-criticism in France's radical political circles had reached not only their own structural scaffolding, but also their moral necessity. Yet as the chapter insertion dates in his 1982 *Théorie du sujet* indicate, Badiou never entirely suspended his research into balancing out an inventive political alternative to the logic of France's centralized *representative* democracy with a philosophical theory on the mathematical *presentation* of manifolds. Mathematical philosophy had for him merely been inching silently forward through the murmurs of research.

When the time did come for a full-scale readjustment of philosophical history beneath this research program, a masterpiece bore out the expectations. To the acclaim of an astonished, almost weary philosophical community, *L'Être et l'événement* was published in 1988. Few works of such stature and scope have been published in France since. With *Manifeste pour la philosophie* (1989), *Le Nombre et les nombres* (1990), and *Conditions* (1992), Badiou established his philosophy as among the most important his generation was producing. He broke loose in the field of mathematical philosophy, which had until then sunk into the background of the structuralism and hermeneutics practiced in France. Moreover, the absence of mathematics in the reception given by the Anglophone world to French philosophy would end up tainting the latter's progression through the 1980s. Its English-language double wound up replete with the *idées fixes* that came to be known as French postmodern thought.

Given the scarcity of English translations of Badiou's work until most recently, this mathematical background has been but hearsay. Along a different line, journal readers had become familiar with his essay on Samuel Beckett.[3] Its publication predated the translation of *Manifeste pour la philosophie* and two other essays from *Conditions,* which were published by the present author finally in 1999 as *Manifesto for philosophy.* Still, with respect to the specific orientation of the works cited above, these pages were philosophical immersions in a historical mode. The translation of Badiou's hotly debated and fiercely contested work on Gilles Deleuze soon followed. Then came the translation, appearing in 2001, of a work that had marked another turn in Badiou's ventures temporarily away from mathematical philosophy, *Ethics. An Essay on the Understanding of Evil*—originally published in France in 1995.[4]

Since then, several translation projects of the mathematical works have been organized, but delayed in their long course to reach the minds of readers. Badiou's clout has risen, as have the costs of translating him. The present translation itself was first proposed to State University of New York Press and the Editions du Seuil early in 2001.

Corresponding to the line of his French output, then, what we have in English of the original texts is the Badiou of his post-mathematical rush of the late 1980s. Indeed, we have a thinker whose deep roots in the fields of contemporary research are squared off against the France of Althusser, Lacan, Mallarmé, and Beckett. As such, Badiou's scope is altogether recognizable as integrating the traits of French thought. Where does the difference then lie?

In fact, Badiou's work culminates the strong neo-positivist current that had developed on the margins of Althusser and Lacan's seminars in the *Cercle d'Épistémologie de l'E.N.S.* (the École normale supérieure Epistemology Circle) in Paris in the mid-to-late 1960s. The Circle's main publication organ was the *Cahiers pour l'Analyse*. In English, its contributors' collective input can be read in the celebrated introduction to Michel Foucault's *Archaeology of Knowledge,* although it must be said that the *Cahiers* is deserving of a broad translation proposal. As Badiou swept up several *succès de librairie* in the 1990s, philosophical best sellers written in the essay mode, and a series of plays for the theatrical stage, his mathematical output again subsided. Meanwhile at the research level, he kept refining his thesis on mathematics, that is, ontology, during his Saturday afternoon doctoral seminars held at the Mathematics Department of Université Paris VI Jussieu.

The program took a notable turn after a critique and suggestions made by Jean-Toussaint Desanti. They were delivered orally during Badiou's Habilitation in 1990 and in print for the special *Les Temps modernes* edition on *L'Être et l'événement,* which consolidated his philosophical standing in France.[5] Schematically put, Desanti offered a hedge to Badiou to broaden his commitment to mathematics within the scope of what he referred to as an "intrinsic ontology" by including category theory into his argument and system.[6]

The work translated here is a major step toward tightening the knot in Badiou's transformation of Desanti's wager in the way it unfolded in France itself. Originally published in 1998 under the classical French title of *Court traité d'ontologie transitoire,* the book is part of a simultaneously published trilogy. Its other two panels consisted of a work on aesthetics, *Petit manuel d'inesthétique,* and of installments in his project for an emancipatory politics of invention, *Abrégé de Métapolitique.*

My choice of a main title, *Briefings on Existence,* refers specifically to the content of the text and core of its argument. Furthermore, it appears to me to ring better in a consumerist market that unfortunately has not left the academic press unscathed. Badiou's reworking of ontology, taken in its Aristotelian sense of a science of being qua being, investigates how foundational mathematics conceptualizes existence, and how this plays off against Heidegger's groundbreaking distinction between Being and beings/existents, or the ontico-ontogical difference—the fundamental difference between the "there is" and "what exists."

The type of existence eventually discussed in the book may be more familiar to English-speaking readers as "mathematical existence." I mention this for the sake of continuity, though I prefer not to hasten the use of such terminology. At any rate, its connotations will be debated by Badiou in due course. Let me simply refer to a similar albeit divergent perspective from P. Benaceraff and H. Putnam. They themselves concede to "leave open the question of whether mathematical existence is a different sort of existence or the existence of a different sort of thing, or both, or neither."[7] In my view, the sense of existence developed by Badiou in the present book is best captured in the implications stemming from Georg Cantor's discovery of sets and their relation to the infinite, which Badiou ascertains as standing at the crux of mathematics and logic, as well as philosophy. As he puts it, Cantor is the one who "fully recognized not only the existence of infinite sets, but also the existence of infinitely many such sets. This infinity is itself absolutely open-ended. It is sealed only by the impossible and therefore real point that makes it inconsistent, namely, knowing that the set of all sets cannot exist."[8]

Badiou goes on to add that Cantor's insights on the infinite actually accomplish the Roman philosopher-poet, Lucretius', a-cosmic philosophy. His was a solidly materialistic enterprise whose concepts rendered in poetic form would set the groundwork for the transformative emergence of modern science, once its postulated predecessors had been mathematized and their objects *proven* to exist. But what exactly the mathematical tools are which best deal with existence, without putting the whole edifice at risk by making it vulnerable to the vagueness of belief or to the idealism of an existing "substance," is the flesh and bones of the inquiry here presented as a "transitory ontology."

Transitory is its nature insofar as we are in the midst of Dasein, being-there. But this being is not motivated by the "existentials" in a being-for-death. Rather, Badiou hovers over Aristotelian terminology to isolate the tension found between the senses of "existent" (*on*) and Being (*einai*). Aristotle famously wrote "the term 'being' is used in many senses, but all of these related to something which is one and a single nature (*pros hen kai mian tina physin*), yet not by homonymy."[9] What does Aristotle mean by saying the senses of being are the same but that the notion of "relation" expressed in "*pros* hen" is more adequate to them than homonymy? In his detailed exegesis of the senses of being in Aristotle, Pierre Aubenque notes that "the doctrine of the *pros hen legomenon* [that the different senses of being refer to a single term] perhaps grounds the unity of Being, but this unity remains problematic."[10] In other words, "it is paradoxically legitimate to speak of being qua being at the moment when the ambiguity of this expression of homonymy is acknowledged."[11] As Aubenque went on to show, Aristotle had to craft a new sense to the notion of homonymy based on the idea that nature, instead of rhetoric, logic, and the categories, would provide a sense to the term "same."

By assimilating Dasein to a *pros hen,* Badiou sets this natural being as a never-ending "toward Being-ness." It is a thought on being that conceives of Being in terms of One as if in a rippling shadow through which only the void actually is. From this void emerges the eventful breakthroughs triggering a specific subjective orientation whose task will be to understand the event's being, the truth of its being. Faced with the Aristotelian analysis of the homonymy of being, Badiou does not hesitate to place the problem with the notion of "relation" per se, namely, in the way that Aristotle had established "relation" as primary to the Platonic doctrine of mathematical objects. In Badiou's view, this is why a doctrine as profoundly Aristotelian as category theory cannot provide mathematics with a foundation adequately dealing with the concept of the One and the Void, within whose tension Set Theory inscribes what Being is. Such grounding can only emerge from an ontology that is *intrinsic*—and thus not based on the Relation first.

Badiou's major work on mathematical philosophy came about precisely through an ontological analysis of the notion of the One within the framework of Set Theory. *L'Être et l'événement* (to be published as *Being and Event*) argues massively that ontology, insofar as it is the science of Being qua Being, is actually mathematics. It is the "mathematics" of Set Theory, which Badiou takes as the foundational basis of mathematics per se, of "what there is." Conversely, mathematics is ontology in terms of what it *says* of Being qua Being. Badiou repeats his assertion in similar form in the pages herein: "Whatever can be said rationally of Being *qua* Being, of Being as deprived of all qualities or predicates other than the simple fact of being exposed to thought as an existent, is said—or rather written—as pure mathematics."[12]

Among the things mathematics does say is that a set allows two quantifiers, universal and existential. Like in Heidegger, the universal extension has no absolute form. If ever its boundaries were broad, as with large cardinals, existence would not be one of its most evident attributes. Were Russell's paradox taken as a wager, claims for the existence of a set of all sets would not even be desirable at all cost. Even when attributed as a second-order property, as in Frege's enterprise, existence exerts a high toll on a rational system intent on formalizing the nature of mathematical signs and arithmetical reason.

A set joins up explicitly in Badiou's reconstruction of Set Theory as the philosophical concept of multiple. The immediate question is: from what tradition does the reading of the multiple draw? For were one to take the "multiple" as a form of the "many" from Plato's ontology, then clearly, "multiple" has nothing to do with Set Theory (i.e. ZFC: Zermelo-Fraenkel Set Theory with the axiom of choice). That is because the set of all sets does not exist— which is not the case for Plato's conception of the multiple. In his epistemology, the multiple is the set of copies subsumed under the One, with the One being in this case the set of all multiple sets. By contrast, Badiou clarifies the issue descriptively when saying:

> What ontology tells us in the theory of the pure multiple is that, inasmuch as a multiple exists, we can only declare its existence inasmuch as it belongs to another multiple. To exist as a multiple is always to belong to a multiplicity. To exist is to be an element *of*. There is no other possible predicate of existence as such. The immediate consequence

is that to exist is to be in a situation, without needing to fall
back on the transcendental, since it is a law of being.[13]

Therefore the sense of multiple has been edged free from a Platonic ontology, though not because existence would require another
term by which to be localized in a proof. Besides, Badiou has explicitly
declared his philosophical intention to think the manifold as cast in a
"Platonic gesture."[14] However, he has taken the mathematical factor in
relation to Plato's essentialism down the road of work done by Gilles
Deleuze. For Deleuze was the one to most systematically undo
Platonism's central dogma by which the Many/Multiple lies in perpetual subsumption, and indeed submission, to the One.

Deleuze's conception of *simulacrum,* crafted in the wake of
his critique of Plato, made way for a broadened conception of the
multiple in French philosophy. It worked as a prototype along a
triple front—epistemological, aesthetic, and political.[15] Human aggregate models incorporate nonhuman forces by which to understand what it means to be collective, that is, *many.* The *simulacrum*
is the prototypic agent for the stimuli of which rhizomatic diagrams
would later appear in Deleuze's work—all of which are crafted
from the notion of multiplicity and manifold.

Just as Deleuze points out that the "reversal" of Platonism was
already prepared by Plato himself as he vouched for the originality
of a model that would define the Sophist, so also does Badiou
reintroduce Plato as the philosopher to have first ascertained the
four conditions of philosophy from its production of modeless copies: the matheme, poem, love, and political invention. The point to
bear in mind here is that in Badiou's view these conditions must
coexist in the sensible realm for philosophy to arise, and coexist
immanently if it were really necessary to stress this point against
Deleuze's claim whereby some "transcendence" is at work here.[16]
These two instances converge into supplementary prescriptions. Their
upshot is that Plato would have thereby initiated the factor of decision making within a philosopher's relationship with the system
he recognizes as *necessarily* unfolding through his thoughts. There
may not be much talk on liberty in Plato, but there is most defi-
nitely space to decide and prescribe.

With such space made available, any philosopher is bound to
display the failings of human decision making—"human error"—

when it comes to riding the wave that unfolds the complexities of an aggregate conceptual doctrine. Just as Plato provided nocturnal prescriptions to deal with political dissent in a troubling scenario in which, by all appearances, Socrates himself would have succumbed, so also did he intensify the subsumption of the many under the One by banning images that did not adhere to Idea-types from the Polis and the scope of his philosophical inquiry. From Plato's writings the question nonetheless emerges: what type of decision making is most appropriate to a philosophical gesture vying to maintain the rights of the many to emancipation?

Badiou's singular contribution rests with the broken symmetry he sees as crosscutting two conceptual worlds, formerly Plato's intelligible and sensible realms, as well as the irreversibility of the decision on the methodological approach to take when dealing with this broken symmetry. According to Badiou, "it is when you decide upon what exists that you bind your thought to Being."[17] With respect to ontology, Badiou sees these worlds as incommensurable in relation to how the manifold is conceived. His contribution lies in sketching out the need, for the sake of a consistent ontology, to choose between the two and separate them. He thereby makes choice the moment of a metaontological decision between an intrinsic ontology, in which sets are real and endowed with properties, and an extrinsic one in which relations occupy the fore. In that framework, allotting existence with its appropriate status is as important as the question of truth itself. As Badiou writes:

> Due to its decisional dimension, mathematics can only become prey to the question of its norm. More accurately, it is prey to the question of the norm of what thought is able to sustain as an assertion of existence. Do we have to confer existence on numbers whose principle is to no longer consist of units? Must existence be conferred to nondenumerable actual infinite sets? What conditions can ensure a well-formed concept to accept an identifiable extension? How are the assertion of existence and the construction protocol linked? Can the existence of intelligible configurations be granted despite the impossibility of displaying even a single case of them? These questions will be settled according to an immanent norm that does not constitute thought, but orients it.[18]

The norm Badiou seeks will be one to have always maintained the manifold in subtraction from the One. Indeed in Badiou's work, the One is either in excess or subtraction. It not longer dominates over Being; it no longer "is." This is about as full circle as anyone is able to go in redesigning Plato.

What unleashed the need to liberate the multiple from its boundaries in Badiou's thought was the legacy of German Romanticism's conception of the infinite. At a time when the mathematical groundwork behind the future Set Theory had unbound the infinite from the One and the Absolute, the romantic exaltation of the infinite seemed to come to a halt with Nietzsche's critique of God. Yet even in a negative mode, late Romanticism still harbored the infinite under the sign of the sacred.

Accordingly, Badiou begins *Briefings on Existence* with a meditation on the death of God that deviates from Nietzsche's path by *multiplying* the gods once exalted within philosophical discourse under a single proper name. Multiplication is the key operator here. It is only fitting, then, after the long discussion in chapter 5 on the finite and infinite intellects in Spinoza's philosophy, to put forward the neo-Cartesian idea according to which "God must be understood as mathematicity itself."[19] The discussion asks more precisely which of the gods were said to exist and how.

In his own view, Badiou is led to the continued multiplication and numeration of things metaphysical through Georg Cantor's double breakthrough. Trivially, Cantor's discovery of cardinal numbers removed the top from number chains. But what worked more specifically for vaster denumerability was his Continuum Hypothesis. The three orientations in thought fostered in *L'Être et l'événement* and *Manifesto for Philosophy* are based on the Continuum Hypothesis' undecidability. Descriptively, the Continuum Hypothesis asserts that the number (or "power") of members of a set S is equal or less than the number (or "power") of subsets of a set S. Given that in Set Theory every member is a multiple, and that every multiple is transversal to itself, a subset would be a multiple of a multiple, that is, a second-order multiple. The upshot is that these aggregate sets within a set outnumber or are equal to the set of individual members within the same set, which are the "situation states" and "situations" in Badiou's terminology, respectively. But an aggregate, typically, also has emergent properties, which puts its own features partly outside of the reach of its individual components' properties. To what extent and degree? This remains part of the dilemma in the undecidability of the Continuum Hypothesis.

The three orientations portrayed in *L'Être et l'événement* dealt with the possibilities of minimum exceedance, a maximum peg and a floating determiner. Minimum exceedance refers to the shortest gap possible between the denumerable chain and the infinite aggregates. In ontological terms, one could locate this orientation as a case in which the power of the continuum is equal to the Aleph immediately succeeding it. That is, the Continuum Hypothesis flows, linearly we might add, from ZF with the axiom of constructibility, which is the formalized version of this conclusion. Not coincidentally, this evokes an ancient Greek choice. For it was Gödel, an illustrious Platonist, who proved the set-theoretic determination in the axiom of constructibility of sets. Essentially put in Gödel's 1929 Completeness theorem, the set-theoretic universe is equal to the constructible universe (V=L). In Badiou's terminology, this orientation amounts to postulating that a situation state always follows the situation itself, thereby allowing linguistic sense as *representation* to control Being's *presentational* exceedance.

The maximum peg, on the other hand, is the transcendent disposition par excellence, in which the infinite is welded to the Absolute-One. Here we have a rendition of Leibniz's attempt to eliminate the void from his system, and link up the infinitely small realm of the *petites perceptions insensibles* with the infinitely large reserve and contours of the universe's objects. His formula $1/\infty$ anticipates Cantor's understanding of the Continuum as a rational body of continuity. But it also sets a theological cap upon any emergence of events that derail the particular blend between sense and truth that the theologically inclined are wont to forge under the name of "faith." This orientation tends to seal off the universe from outer deviation, for plainly there is no outside in which to roam.

Roaming, however, is what the third "generic" orientation is all about. Badiou edges along its possibilities as if on a line of flight by which to keep a view of ontology as a science operative as a system. But this is a system of holes drilled through it—a phenomenon he terms as describing the truth's relation to sense. In these terms, not only does truth not make sense, but it is senseless. It does not make sense (at least not always) insofar as truth is what is produced before being ascertained in the infinite movement within a given situation in each of philosophy's conditions. It stands without sense insofar as sense is not what legitimizes a truth. Instead a truth is legitimized by the fidelity a Subject displays to the event creating it through the recognition of its own vocation as a task to know the event's emergence in terms of truth.

These conditions are crossed by a line stemming from the event that immanently constitutes a rational body, which Badiou conceives in the elementary terms of a system.[20] As the notion of "system" is inherently vulnerable to criticism owing to the logistics of our times whereby any system, if only to prove its scientific value, is sent vacillating, Badiou encounters a major challenge to this application of Set Theory through the ontological reading he asserts of Cantor's Generalized Continuum Hypothesis. Insofar as Set Theory lends itself to Badiou's thesis, it does so due to its grounding function for mathematics as a whole. This function was itself derived from its initial task, namely, to ground the reasoning processes underpinning the operations in number theory and arithmetic.

What then to do when another mathematical approach vies for grounding status? For Badiou the question is to identify its proper localization within the system by following through in its logic, conditions, and possibilities, while doing the utmost to not force its place.

The ambitions and power of Category Theory show just such an approach. Category Theory emerged in the 1930s with the rise of abstract algebra. Two mathematicians, Samuel Eilenberg (d. 1998) and the late Saunders Mac Lane (d. 2005), conceived of several features in structures studied in abstract algebra, namely, homomorphism, substructure, and isomorphism, not only as having universal applicability, but also as appearing to be independent from Set Theory. A category consists of arrows and point-objects that acquire their properties only from the various arrow relations in which they are configured. Unlike the intrinsic dimension of Set Theory, the notion of category was not perceived as most fundamental. Instead, the *functor* was.

The arrow in a transformation thus became more important than the objects the latter links or maps. As Eilenberg and Mac Lane wrote: "it should be observed that the whole concept of a category is essentially an auxiliary one. Our basic concepts are essentially those of a *functor* and of a natural transformation."[21] This is why a category is in fact nothing without its criteria. It is a collection of "objects" satisfying three basic conditions: that any pair of "objects" *a, b,* are linked by a collection of arrows or "morphisms"; for every triple *a, b,* and *c* of objects, the operation of arrows from *a* to *b,* and *b* to *c* is called the "composition" of *a* to *c;* finally, every object a has an arrow linking it to itself called the "identity morphism."

In a nutshell, elementary category theory shows that there is no object prior to the relations constituting it. The theoretical apparatus literally *illustrates* this by using a geometric presentation combined with its equational transcription. It is not merely a matter here of delving further into the question of the meaning of a multiple being transversal and open-ended. This is enough to ward off any claims of essentialism going on behind the scenes. It is not even just a matter of taking Aristotle's typological rebuttal of Plato's numerology as a path by which to undermine the claims emerging from Set Theory and turn them towards a logic of being.

What it is in fact is a metaontological theory, very Deleuzian at that, of showing and analyzing the emergence of (mathematical) objects *qua* forms, that is, structures, in possible worlds. This sense of emergence provides an altogether unexpected necessity to the notion of existence, albeit within the realm of the possible. As J. Bell writes:

> Set theory strips away structure from the ontology of mathematics leaving pluralities of structureless individuals open to the imposition of new structure. Category theory, on the other hand, transcends particular structure, not by doing away with it, but by *generating* it, that is, by producing an *axiomatic general theory of structure.*[22]

Even more significantly, Bell goes on to add that category theory has succeeded in the debate over mathematical foundations since it has been able to show the "ubiquity of structure" in mathematics.[23]

In today's network society, there is scarcely anything astonishing about a mathematical theory placing relations before objects in contexts it calls "possible worlds," or "universes." Yet despite its "interconnectedness" our world still remains crossed by endless convergence on objects, things, and substances, endless reification, endless inertia—regardless of the ties that bind them and us to a situation and situation states. Emerging from French structuralism in the social sciences and a student of structuralism in mathematics, Badiou remains unconvinced, as do any number of mathematicians, of category theory's capacity to provide the grounding that the manifold or pure multiple, defined as a multiordered, dense collection of empty-set aggregates, does through axiomatically derived theorems instead of illustrations. In his stance, one finds the letter still vanquishing in its claims over the image.

As category theory exerts no overall claims on rebutting Set Theory, one can surmise sets to be real mathematical objects. Or with respect to possible world semantics in modal logic, if a set exists in every possible world created by category theory, then a set is said to exist necessarily. Insofar as a set does, the onus has been on category theory to obtain a definition—that is, a "presentation"— of what a set is within its own domain, with its own tools. As C. McLarty points out "set theorists did not consider sets and functions equally important. [. . .] Set theorists thought in terms of a discrete universe of sets linked by functions."[24] Early in the twentieth century, topologists had already conceived of topological space as connected by maps. From that conception emerged the category of "topos," which is category theory's work to formalize and reduce the notion of set within its own boundaries.

Category theory did not merely reconstruct topology within its own scope, it literally absorbed it. As of the 1960s in the work of Alexandre Grothendieck, the study of toposes became a central element of category theory. It was the study not merely of a collection of arrows between objects, but a collection of categories, or "sheaves," between them. A case in point is that a topos is rich enough a concept to integrate most of the mathematics done in undergraduate departments. Accordingly, a sheave provides a topos with a probe-head, that is, a set varying through space. A topos also includes a diagram that may be understood as a set varying through discrete time. These are two basic, but fundamental examples.

This diagrammatic potential stands precisely at odds with Set Theory. The presentational core of internal transversality in the latter, that is, an untotalizable collection without a contour, is not homologous with the representations that can be made of a set, as in Venn diagrams for instance. More interestingly, it also gives a vantage point over Set Theory, although it generally does this in terms of possibility and not necessity. For according to possible world logic, it is simply not the case in all possible worlds that set theory does not provide an axiomatic system by derivation of which theorems describe only a *single* universe.

Admittedly, one of the claims made by category theory is that two key operators of bivalent logic, namely, the excluded middle and the reductio ad absurdum (i.e., wherein the double negation in subordinate clauses entails an affirmation of the original postulate), do not obtain in all possible universes. The capacity to make and check such a global statement in terms of possible worlds suggests that category theory, specifically in the truth evaluations afforded by topos theory, can assess a universe's inner logical possibility from outside, as it were. Ultimately this leads Bell to contend that "topos theory has striking connections with *logic*."[25] This position links up directly with Badiou's, for whom "categorical thinking is not onto-logical. It is essentially logical. It is not a proposal on universes, but an ordered description of possible universes."[26]

I have discussed my own perception elsewhere as to some of the possibly conflicting points in Badiou's localization of category theory within a philosophical lay out and what category theorists themselves have expressed regarding the theory's foundational claims.[27] My intention in that essay was not to suggest that mathematicians should dictate to philosophers how their field may be inspected. After all, Badiou is quite clear in denouncing the split figure mathematicians often provide the public of their art. For the practically minded, numbers are thrown out in dizzying array from computing machines. Yet behind the scenes, as if reserved for the thoughtful, the nature of numbers is kept within the technical density in which most are discouraged from participating.

Despite its basis in the split Subject, does this renewed interest in axiomatics with Plato make Badiou an "essentialist"? The answer to this question is certainly not simple, and perhaps not entirely pertinent by the standards and interests of French philosophy today. Perhaps nowhere more than in chapter 13's discussion on groups can the bounds of inquiry and scope of permissibility in French philosophical problems be felt.

From the outset of the *Cercle d'épistémologie*'s work published in the *Cahiers pour l'Analyse,* the space for thought which opened in French philosophy stemmed directly from the two strands of Descartes' oeuvre: mathematics and the Subject. Many French thinkers recognize the Subject to be a French affair. This partly explains their affiliation to thinking it through as a philosophical constant—regardless of, and consequent to, the deconstructions it has suffered. Then again, the mind/body affair is equally Cartesian. But the transhistorical fidelity in the Anglo-American sphere to T. H. Huxley's biomechanical translation of Descartes' substantial dualism by an elision of the complexities of Subject keeps parting philosophy's cultural divides. The fallout from Wittgenstein's private-language argument unwound in Kripke's hands has considerably slid this issue over to questions related to Subject—without, for all that, the name of the latter becoming a necessity.

In turn, Badiou continues the work on Subject within psychoanalytic theory in a tradition that conceives the latter as the parallel space of thought to the Cartesian mind/body issue. In his system, psychoanalysis "thinks the great categories unpinning the subjective tress: the letter, the identity of inertia, the meshwork of the Same by the Other, the operatory plurality of the identical, the image . . . "[28] All of these categories were first set out by Plato in the *Sophist* as the five genera, then perpetuated by Descartes as he worked to establish an element of certainty to existence, and most recently split into a kaleidoscope by Lacan's reworking of Freud's first topos.

Adding even more to the complexity of answering the question on whether Badiou is or is not an essentialist is that he deliberately undermines the irreducible nature of the split between essentialists and process thinkers. Part of his move reiterates what he sought to prove as driving Deleuze's ontology. Readers of Badiou's *Deleuze* will recall the surprising shifts to which he subjects Deleuze's thought: the process thinker becomes the proclaimer of Being's univocality; the orgiastic "libinal" sensualist becomes a ponderous meditator of death; the free, anarchic advocate of lines of flight becomes the Stoic who is motivated by taste instead of necessity. Capping things off, we discover the nomadic cutter of multiplicities to be the front office urban clerk version of a gatekeeper to the One. These are no mere claims, nor are their supporting references taken out of context. Certainly, this is not how Deleuzians usually represent themselves. But that was precisely the point.

Badiou raises the question of the conditions required to think along Deleuzian lines and whether they are consistent beyond their affirmations. By extension, he shows how nonessentialist thinkers cannot so easily reach satiation regarding their poetic, figural, and metaphoric innovations—no matter how much creation is displaced in favor of simulation. The task of shedding the essentialist paradigms, at least as far as philosophy may be concerned, is perhaps subject to greater caution when the analysis shifts to a mathematized field, or to mathematics itself. If anything, this is one of Badiou's leading intuitions.

As mentioned previously, from the outset of his ontological thesis, Badiou has asserted the character of a system for his thought. As he wrote in *Manifesto for Philosophy,* "if by 'systematicity' we understand, as we must do, the prerequisite of a complete configuration of the four generic conditions of philosophy (which, once again, in no way demands that the *results* of these conditions be displayed or even mentioned), by means of an exposition that also exposes its expository rule, then it is of the essence of philosophy to be systematic, and no philosopher has ever doubted this, from Plato to Hegel."[29] The feature of "exposing its expository rule" is essential to the transparency in Badiou's approach. Like any rational scientific system, it thus submits its global claims to the evaluation of readers and reason alike.

On the other hand, one of the main objections raised against Badiou's theory points to a lack of necessity in equating mathematics and ontology. Badiou's conjecture deploys a vast arsenal of proof approaches, from logical and ontological arguments, to the history of ideas. At bottom, criticism on the lack of necessity to his argument is inclined to bypass the ontological sphere itself, which it deems to be nonessential for an effective philosophy today. So be it, philosophy might have numerous facets. When it is called upon by society in a non-hypocritical sense, as it is occasionally nowadays in relation to ethics, it can choose to follow—despite the inner preoccupations of its more rigorous commitments.

The purpose of this introduction was primarily to reconstruct the line of Badiou's introduction to the Anglophone world according to his production in mathematical philosophy, that is, ontology. Badiou's conjectures are ultimately fundamental for establishing metaphysics, philosophy, and ontology, that is, mathematics, as the scientific sphere of philosophical activity—as well as the philo-

sophical sphere of the sciences—which remains their vital component to ward off dogmatic big science big industry coagulations, or
the top-down forcing with which the current American Administra-
tion is pressuring its scientific community.

   Let us take leave, then, of Badiou's ontological argument here
as it complexifies so as to allow him the task of unfolding both the
need to integrate category theory into his system as well as the
necessity to situate it as a logic of being—as an "onto-logic."

   A few observations are due regarding the English version of
French translations of foreign language texts found in the book.
Badiou is a versatile translator from the Greek and Latin. This
means that he also brings his own reading to Spinoza, Plato and
Aristotle, for example. Despite the advantages of including the clear
prose of G. H. R. Parkinson's translation, which is my basis for
citations from *The Ethics*, I have had to adjust or indeed translate
directly from Badiou's own adjusted translations from any number
of prestigious French versions in order to keep the citations coherent with his arguments.

   Likewise, Badiou redesigns many of the French translations
from the Greek, which are already different enough in detail to the
English translations. I have chosen to translate Badiou directly, then.
When appropriate I do offer a standard English translation of the
original. This is not an entirely comfortable choice for a translator
whose job, after all, is to bring the text into another language in an
accurate rendition that manages to filter out the sounds and background hiss of the original linguistic environment. Deliberately al-
tering translations of classics in English to suit an author's own
alterations is somewhat like being complicit to a not entirely kosher
practice according to the ethics of translating. However, Badiou
interprets through his own translation in order to highlight and
emphasize what in his view is deserving of conceptual notice. It is
this commitment to which I have devoted myself in this translation.

   Needless to say, a formalized theory of interpretation is sorely
lacking in Badiou's work. The scholarly guidelines authorizing his
translation decisions, for example, are inexistent. On the other hand,
Badiou's philosophy is based on decisions and prescriptions. It is a
militant philosophy, which he wears on his sleeve. Indeed, it is
another universe from the public ethics we commonly find today in

which journalists or intelligence agencies, let alone accountants and auditors, deliberately alter reality in order to suit the facts, all other things being equal.

If the postmodern was characterized as an age of pastiche, all of us, regardless of our political leanings, may agree that we are now in the age of plagiarism and fraud. The Washington D.C. neo-Straussians have ushered intellectuals to the brink of the nihilistic naiveté of the postmodern age. What stands out is that not only is postmodernism over, but we also need not be convinced that the only way out is forward. The task of analysis involves exposing the basis of a theory's presupposed reconstruction as a method by which to think through the time of scientific—and not merely technological—progress. The social, political, and now cultural and scientific records of North America are beginning to prove the point that historical knowledge does not quite follow a line, be it a mathematical line.

There is no phenomenon in France akin to the conquest of analytic philosophy over epistemology and the philosophy of science as occurred in the Anglo-American sphere. Moreover, the French scientific community has benefited from a secular education steeped in philosophy more than America has. A critical philosopher can only be astonished to what degree religion, and its bastard son, "self-help," keeps displacing philosophy once the limits of reason go beyond the practical. Yet again the limits of empiricism prove to be the limits of technology.

For this translation, I would like to thank my wife, Monica, for her love and guidance in life as in philosophy. Alain Badiou granted me the rights to translating the work in early 2001 and kindly provided the preface. And I would especially like to thank James Peltz, director of State University of New York Press, for his patience and perseverance in the lengthy elaboration of this project.

PROLOGUE

God is Dead

What is God the name of in the formula "God is dead"? We cannot assume that *for us* there is anything self-evident about this matter. Besides were God indeed dead, then, as happens to the deceased whose very tomb is no longer but an erased or muddy stone, it is plausible that the memory of what this name "God" was all about is likely to be shrouded, dispersed, and deserted. Moreover, there is a complete difference between the theoretical formula, "God does not exist," and the historical or factual statement, "God is dead." The former is akin to a theorem asserting that no rational number exists such that it can create a relation between the side of a square and its diagonal. What's more, it presupposes that God is a concept whose meaning is reactivated by its unceasingly demonstrable theorem of nonexistence. By contrast, to utter "God is dead" makes God into a name, a proper name. It is similar, then, to the way that great-great-grandfather Casimir Dubois is said to be dead, perhaps without knowing anything or understanding anything—save for his death—about the distinctive living infinity arranged under the phrase "Casimir Dubois".

The question is even more acute in that it assumes that for God to be dead it must certainly be argued that He has been that way for a long time. Perhaps was it right after Saint Paul's sermon that people started making God die in His only real life, that is, Christ's resurrection? This was a unique and decisive victory over death, death as a figure of the Subject, and not of biological objectivity. At any rate, onto the living God the Renaissance

21

superimposed the suspicious multiplicity of mythological gods. About them Ernst Renan would much later go on to say that they were so present and naked in great classical art as to all be wrapped in the purple shroud in which dead gods sleep. And had it not been for the Renaissance, Galileo or Descartes would have introduced the universe as a type of material graph of mathematicity. They froze God, as it were, in the trans-mathematical punctuality of actual infinity— which is but to live a "literal" death. Then again, it would have been the philosophers of the Enlightenment for whom politics is strictly the affair of humankind, an immanent practice from which recourse to the All Mighty's providential organization had to be discarded. Or, Cantor would have been the one to chase God from his infinite setting. In His place number and computation would have been installed. It can well be that God has been agonizing for a very long time. What is surely less doubtful is how, for centuries, we have been busy with successive ways of embalming Him.

This is why the question of what lies in his name is increasingly obscure. It is not by fine-tuning its function to that of the Father that we shall be enlightened. Feuerbach had already asserted that the Christian God, with all of its fixtures, was only the projection of family organization and its constituent symbolism.[1] But he could only do so because God was already dead, or dying. Let us say that this thesis on God participated in the process of His death and the oblivion of what was living in His name. Bringing the way God functions back to the Name-of-the-Father is but one amongst many ways of effacing the living God's singularity behind its own name and the ideal of science.

This is the crucial point regarding the formula "God is dead." Simple and difficult at the same time, it can be said as follows. If "God is dead" is asserted, it is because the God spoken of was alive and belonged to the dimension of life. When you consider a concept, a symbol, a signifying function, you can say that they have become obsolete, contradicted, and inefficient. They cannot be said to have died. This is why grasping the question of God beneath the heading of some primordial symbolism ends up by concluding that He is not dead, or even that He is immortal.

From this standpoint, psychoanalysis has kept an ultimately ambiguous relationship with the question of God. To the extent that psychoanalysis has done so along the lines of Feuerbach, that is, by the sublime name of one of the functions whereupon "desire" is pinned, it pursues science's mortification of transcendence. But to the extent that psychoanalysis actually situates this function's stability, and that no subjective constitution can radically do without it, it guarantees God a conceptual and unprecedented durability from within His own supposed death. As empirical proof, observe that a number of eminent and talented psychoanalysts, such as Françoise Dolto and Michel de Certeau, saw no formal contradiction between their Freudian involvement and religious belief—quite the opposite. Not anymore than did Lacan, who can certainly not be suspected of clerical complacency. Yet he contended that it is strictly impossible to finish up with religion once and for all.

Now, my conviction on this matter is the opposite. I take the formula "God is dead" literally. It *has* happened. Or, as Rimbaud said, it has passed. God is finished. And religion is finished, too. As Jean-Luc Nancy has strongly stated, there is something irreversible here. What is ultimately important in this is to figure out the subjective mechanism explaining how people can so easily believe that it is nothing of the sort and that religion prospers; or even, as it is so often said at this time, that religion returns. Admittedly, nothing returns, and we do not have to believe in specters. The Deceased drifts away solitary and forgotten in His anonymous, stateless tomb.

Provided, of course, one posits that only a living God could have died without a possible resurrection. What is a living God? A living God, like all living beings, is whomever other living beings have to live with. This is what Blaise Pascal, one of the last defenders of that condemned God, had understood admirably. The God who can die is not, and cannot be, Descartes' conceptual God, that is, the champion of infinity, the suture of mathematical truths to their being, or a guarantee of judgments in the shape of the Other. It must be Isaac, Abraham and Jacob's God, or the Christ who speaks directly to Pascal in his inner Garden of Gethsemane. The living God is always *somebody's* God. The God with whom somebody—Isaac, Jacob, Paul, or Pascal—shares the power of living in the pure present of His subjective unfolding. Only this living God nourishes a properly religious conviction. The Subject must deal with Him as with an experienced power in the present. He must be *encountered,* and encountered on one's own.

So is Kierkegaard's God just another belated defender of the dying God. When purifying the affect of hopelessness, Kierkegaard states that "in its relation to itself, in wanting to be itself, the Self plunges through its own transparency into the power that placed it."[2] God is dead means that He is no longer the living being who can be encountered when existence breaks the ice of its own transparency. That so and so declares to the press that He was encountered under a tree, or in a provincial chapel, changes nothing. For we know that from such an encounter no thought can use its rights to advantage any longer, let alone do we grant someone claiming to see specters more than the positive consideration of a symptomatic manifestation.

It is in this sense that we must declare the fact of religion dead. When religion does show itself in the apparent range of its powers, it does so only as a particular symptom of a commemoration in which death is omnipresent. What subsists is no longer religion, but its theater. For it is only in drama, as in *Hamlet,* that specters cast a semblance of efficacy. What this ultimately bloody drama represents to us is something we imagine religion could have been, provided the living God—about whom nobody has the faintest idea—were not dead.

Common objections to the motif of the living God's real death, and thus religion's very own, feeds on two sources. These are, on the one hand, the doctrine of sense, and on the other, what are called "forms of fundamentalism," entrusted as they are with conveying the conviction of religion's return.

I do not believe these objections are relevant.

There is no question that one of the functions of religion is to give sense to life, and more particularly to the shadow that life bears, death itself, which is anchored to the real. It is surely not accurate to claim that all kinds of donation of sense are religious, that is, requiring the living God and thus the God who is historically able to die.

On this point, it is crucial to distinguish what the word "God" names in the formula "God is dead" (a point where the word actually touches upon religion) from what the same word names in speculative metaphysics. It is one of the numerous merits of Quentin Meillassoux's recent thesis to have ascertained in a powerfully original ontological and ethical aim the God of metaphysics to have always been the central gun in the rationalist war machine against the living God of religion.[3]

As Pascal objected to Descartes, only a dead God is really suitable for metaphysics. Only a God that has already died or has been dead from the beginning. This God is one on whom no religion can feed its faith, even if it has tried to somewhat tame sundry spirits infatuated with reason by declaring itself compatible with it. This is something that religion basically is not about. For religious risk is to make God into a living being. We try to live with Him, and by living with Him, produce sense for a total life, including death. By contrast, metaphysical risk only goes as far as to understand the word "God" to mean the convincing consistency of a concept and according to this concept guarantee that truths make sense. The word "God" is an amphibole in that it covers, as a living being, the total sense of life and, insofar as always already dead, the possible sense of truths. With respect to God, it is true that religion is vivifying and metaphysics mortifying.

The great metaphysical work of the mortification of God begins with a blast as early as the Greeks. Admittedly, it is marshaled according to sense, to the donation of sense, or to the totalization of sense. But, counter to Kierkegaard the "antiphilosopher," it holds back on all affects and existential plunging into this donation.

In this respect, Aristotle's God is exemplary. Taken from the vantage point of physics, God is the ultimate sense of movement as the supreme immobile mover. But who says life can ever be at eternal rest? This is the very definition of death. All the more so since Aristotle's God moves all things not out of interested action or subjective intercourse, but by the finalized attraction of His supereminence. With respect to the things he moves, this God remains completely indifferent. Who can declare this indifferent and immobile eternity to be living?

Now taking this same God—were it not a question of another one, who knows?—from the vantage point of metaphysics, it would then be manifest that as pure act His only possible office is to think Himself. This God has no acceptable reason to think anything other than His own purity. Here again, there is certainly donation of sense, because it is only by assuming an agent intellect that, detached from matter and supremely turned upon its own perfection, the theory of substance can be completed as an enigmatically singular compound of matter and form, or of act and power. For the principle of singularization—which is act or form itself—must in the end be delivered as an absolute singularity, an act exhausted in its act, or an integrally separated form. The word "God" names these operations of completion.

Moreover, this completion carefully organizes sense under the guise of a demonstration, that is, the existence of God, which is entirely opposite to making attestations on his life. Let it be said then that the God of metaphysics makes sense of existing according to a proof, while the God of religion makes sense of living according to an encounter.

The upshot of this is that the death of the God of religion leaves the question of the destiny of the God of metaphysics unresolved. The latter maintains a relationship neither with life nor death. From life's standpoint, and therefore religion's, this means that He is perfectly dead. The upshot is also that, as far as sense is concerned, the irreducible feature of its effect can be satisfied with operations from which all subjective assumptions of a living God are radically excluded. Besides, this is why Heidegger cannot equate Nietzsche's statement, "God is dead," anymore than the imprecations and anathema against Christendom, with the subsequent deconstruction of the metaphysical schema of onto-theology. These are indeed two distinct questions. A great deal is needed for the disappearance of the religious as such to be mechanically swept away along with the disappearance of the infinitely more resilient metaphysical motif. For the metaphysical God, which names only a first principle, is inaccessible to death.

Therefore, it cannot be argued that the sole workings of sense stand as an objection to the irreversibility of God's death. As metaphysics bears witness to it from its very beginnings, there is nonliving sense, literal sense, argumentative sense and eventually mathematical sense. Such sense deeply severs the religious allocation of sense from the disposition of the living God.

As for contemporary strains of *fundamentalism*, I would argue that one gains nothing from picturing them in terms of a return of the religious. They are contemporary formations, the political and state-based phenomena of our times. Let us say that they are inventions. It has been noticed for a long time how totally unproductive they are on the strictly religious level. The varieties of religious fundamentalism are virulent in the domain they assign to themselves, which is the conquest of power. What is conventionally called religious fundamentalism must really be thought of as a subjective form. I would go as far to say that what is precisely asserted in them is one of any number of subjective types wherein God is dead. This type corresponds to what I call the "obscure Subject" because the truth statement of which God is the application is only active insofar as to be barred, burrowed, or unconscious. Thereupon there are no resources other than to mortify that which constitutes God. This is something about which no psychoanalyst should be surprised. Hence, the desperate and bloody affirmation of an artificial and mortified religion, whose subjectively burrowed real principle is, from end to end, that God is dead. The dramatization of this hidden evidence is given simultaneously in the ceaseless reproduction of this death, under the type of death of those presumably guilty of His death. It is also given in the exasperation of the public make-believe of rituals and body markings, which have only ever been meager performances for the precariousness of the living God.

So as to not miss any crucial points, I should add that between the prescription of the female body hidden behind veils and that of the commercially circulating body, or as Guyotat would say, of the *prostitutional* capitalist body, we deal with the same question. Since dead is the God beneath whose living eye all bodies were exposed and under whose rule portions of the visible were distributed, who can tell us now what must remain hidden? Regarding the female body, which when taken in its entirety is the Phallus itself, one can respond with a maximum or minimum of hiding without ever satisfying what is asked of it. In the deleterious climate of the simulacrum of religious war, in which any true religion is absent, it must be avowed that the death of God can be symmetrically read in the short see-through tailoring of Western dresses—nothing but an epiphany of erogenous zones—as much as in the thick black veils from which nothing shines but eyes. This Subject is obscure at any rate. But the real fact—and there is no way out of this—is that the living God is truly dead.

It does not follow, as I have said, that the God of metaphysics is also dead. On this matter, one must begin with what I should like to call Heidegger's aporia. How can it be that the thinker who determined metaphysics as onto-theology, an overshadowing of the question of Being by that of the Supreme Being, ends up saying in his testament that only a god can save us?[4] Obviously it is only possible if, once again, the word "god" works equivocally. The God who alone can save us is not the First-Principle God of primary concern to the oblivion of Being in Western metaphysics. We should also be able to agree that this god could not be the living God of religions either. For regarding the latter, Nietzsche and, more artfully, Heidegger ratified Its death. In addition to the historically dead God of religions and the subsequently deconstructed metaphysical God, which besides in post-Cartesian humanism can take on the name of Man, one ought to propose to thought a third God or divine principle of an altogether other order.

This god, or these gods, or the divine principle do exist. They are the creation of romanticism and distinctly of the poet Hölderlin. This is why I name it the God of poets. It is neither the living Subject of religion, although it is certainly about living close to Him. Nor is it the Principle of metaphysics, although it is all about finding in His proximity the fleeing sense of Totality. It is that from which, for the poet, there is the enchantment of the world. As there is also its loss, which exposes one to idleness. About this God, we can say that It is neither dead nor alive. And It cannot be deconstructed as a tired, saturated, or sedimented concept. The central poetic expression concerning It is as follows: this God has withdrawn and left the world as prey to idleness. The question of the poem is thus that of the retreat of the gods. It coincides neither with the philosophical question of God nor with the religious one.

The poet's task—or as Hölderlin wrote, his courage—is to bear in language the thought of the God that has withdrawn as it is also to conceive the problem of Its return as an open insertion into that of which thought is capable.

Essentially, the relationship to the poetic God is not of the order of mourning, as the obscure relationship to the dead God can often be. Nor is it of the order of critique, or the conceptual defection of totality, as the philosophical relationship to the First-Principle God can be. Strictly speaking, it is a nostalgic relationship. It melancholically envisages a chance to re-enchant the world through the gods' improbable return.

We can henceforth think Heidegger's aporia. If the end of onto-theology must be endured while awaiting redemption from a divine event, the fact is that the deconstruction of metaphysics and assent given to the death of the Christian God has held the chances open for the God of the poem. It is also why all thought is suspended to the dimension of a return as well as to the gesture it can promise. Heidegger makes the Greek gods into the emblem or figure of the God who can return in the German tradition.

I call *contemporary atheism* what breaks with this disposition. It is about no longer entrusting the nostalgic God of the return with the joint balance consisting of the death of the living God and the deconstruction of the metaphysical God. All in all, it is about finishing up with promises.

This atheism confronts us as a task for thought. What still maintains a power of promising nowadays, and the poetic and political disposition of the return of the gods or re-enchantment of the world, is the consensual motif of finitude. That our exposure to being is essentially finite and that we must forever return to our mortal-being, is the fact from which we endure the living God's death only to uphold, under multiple shapes, the indistinct promise of a Sense that has withdrawn, but whose 'come-back' has not been debarred. Even the subjective universe of politics is constantly invested with a melancholic resignation whose heart is tantamount to a vague awaiting for a return of sense—or of lesser nonsense. This is how a unique polity trivially bolsters itself by making believe at election time that it will be different, that it will subtract itself from its own lack of alternatives and its own uniqueness. From this standpoint, Lionel Jospin was the impaired form of the God of poems.

It is thus imperative, so as to be serenely established in the irreversible element of God's death, to finish up with the motif of finitude. Finitude is like the trace of an afterlife in the movement that entrusts the overcoming of the religion-God and the metaphysics-God to the poem-God.

Doubtless, this task partly concerns the destiny of poetry itself. Today the poem's imperative is to conquer its own atheism. From within, it destroys the powers of natural language, nostalgic phraseology, posturing of the promise, or prophetic destination to the Open. The poem does not have to be the melancholic guardian of finitude, nor a cross section of a mystique of silence, nor even the occupation of an improbable threshold. The poem has only to be devoted to the enchantment of what the world is capable of—as

it is. It has only to discern the infinite *"surrection"* of invisible
possibilities up to the impossible itself. Surely, this is what Philippe
Lacoue-Labarthe takes as the 'becoming-prose' of the poem. Surely,
this is what the poetry of Alberto Caeiro, Fernando Pessoa's
heteronym, as Judith Balso understands him, offers us beneath the
sign of a metaphysics without metaphysics. Caeiro declares that he
writes the "prose of verses." As for the gods, Caeiro declares them
to be neither living, nor dead, as either would still be granting them
too much *pathos*. Instead, they are asleep peacefully—maintaining
reciprocal indifference toward us. At any rate, it is poetry itself that
will conduct, and has conducted at least since the beginning of last
century, the task of putting its own God to death.

As for philosophy, the aim is to finish up with the motif of
finitude and its hermeneutical escort. The key point is to unseal the
infinite from its millenary collusion with the One. It is to restitute
the infinite to the banality of manifold-being, as mathematics has
invited us to do since Cantor. For it is as a suture of the infinite
and the One that the supposed transcendence of the metaphysical
God is constructed. It is upon this suture that the surviving intra-
subjective trace feeds. This trace is the one that, once all transcendence has been abandoned, ties us down to the mated themes of
dereliction: 'Being-for-death' and the horror of the real and finite.

So it is with joy that one must receive the fact that the destiny
of every situation is the infinite manifold of sets. Depth cannot ever
develop there, and the homogeneity of the manifold ontologically
prevails over the play of intensities. Consequently, as disconnected
from finitude, we live the infinite as our absolutely placid sojourn.
It may happen that, by chancing upon an event, some truth sweeps
us away along the open-ended infinity of its trajectory. That is when
the search for sense is reduced for us to the sole numbering of this
infinity, or what another of Pessoa's heteronyms, Alvaro de Campos, used to call the "mathematics of Being."[5]

Our times are undoubtedly those of the disappearance of the
gods without return. But this disappearance stems from three distinct processes, for there have been three capital gods, namely, of
religion, metaphysics, and the poets.

Regarding the God of religions, its death must simply be declared. The problem, which is a political one in the final analysis,
is to ward off the disastrous effects wrought by any obscure

subjectivation of this death. The spectacle lies entirely in the dis-connection of the political from the arcana of State power, and the sequential restitution of the political to the pure subjectivity of pre-scription. From then on, the dead God's spectral authority, which can always criminally bind itself to the megalomaniacal scraps of the State, remains without influence and without effect on minds.

Regarding the God of metaphysics, thought must accomplish its course in the infinite. The latter disseminates thought's reserve over the entire expanse of manifolds.

As for the God of poetry, the poem must cleanse language from within by slicing off the agency of loss and return. That is because we have lost nothing and nothing returns. The opportunity of a truth is a supplementation. Something then arises, but arises here without depth, without anywhere else to go.

Committed to the triple destitution of the gods, we, inhabitants of the Earth's infinite sojourn, can assert that everything is here, always here, and that thought's reserve lies in the thoroughly in-formed and firmly declared egalitarian platitude of what befalls upon us here. Here is the place where truths come to be. Here we are infinite. Here nothing is promised to us, only to be faithful to what befalls upon us.

It is this 'here' that a poet, born so far from us in a language closer to us than any other, the Chouvash poet Guennadi Aîgui, celebrated in a song to the glory of what about the here cannot be replaced.[6] Divine guidance is not what this song celebrates, which is why it is called, "Here." It leads us toward understanding that the here is gained once the search for the dead God's shadow anywhere and under any name is renounced. In this song, even the death of man, a transitory configuration of dispersing infinities, can be en-visaged as maintaining and receiving these infinites. I end this pro-logue with his song:

> Here everything answers itself
> In a primordial and high language
> As one part of life answers
> The contiguous indestructible part
>
> Here at the curling extremities
> Of branches in the appeased garden
> We seek not the horrible clot of sap

That resembles the afflicted silhouettes
Embracing a crucifix in the evening of calamity

And we know of no word or sign
That would be higher than another
It is here that we live, here that we are beautiful

And it is here that we trouble the real when being silent
But if our farewells to it are rough
Life is a part of it as well
As if on its own
A novelty inaudible to us

And parting from us
As the reflection in water of a shrub
It shall remain aside
To afterward occupy
Our place

So that the places of men be replaced
Only by the spaces of life
Forever more.

All of the developments to follow, as abstract as they will often appear, should be conceived as in a meditation, in the clearing of God's death, of what must be thought in the word: "here."

1

THE QUESTION OF BEING TODAY

There is no doubt that the rearrangement of philosophy according to the question of Being is due to Martin Heidegger. We owe Heidegger credit for having named the era in which this question has been forgotten. The history of this forgetting, begun as early as Plato, *is* the history of philosophy itself.

What for Heidegger is the distinctive feature of metaphysics, that is, metaphysics conceived as the history of Being in its withdrawal? We know that the Platonic gesture placed *aletheia* under the yoke of *idea*: the cross section of the Idea or Form as a singular presence of the thinkable ascertains that beings predominate over the initial or inaugural movement of Being's unconcealedness. The upshot is that non-veiling or disclosure is assigned to securing presence. Most important is that this securing exposes the Being of beings to the resources of counting and of "counting-as-one." That by which 'what-is is what is' *is* also that according to which it is *one*. The norm of the thinkable is the unification of a singular being under the power of the One. This norm, this normative power of the One, is what crosses out the "coming-to-itself" or the "re-entry within itself" of Being as *physis*. The theme of *quidditas*, whatness, as a determination of the Being of beings according to the unity of its *quid*, is what seals Being's entry into a normative power that is strictly metaphysical.[1] It destines Being to the preeminence of beings.

Heidegger summarized this movement in the notes included at the end of volume II of his *Nietzsche,* which he titled "Projects for the History of Being as Metaphysics":

> The preeminence of *quidditas* continually brings about the preeminence of beings themselves in what they are. The preeminence of beings secures Being as *koinon* (common)

33

from the perspective of the *en* (One). The distinctive character of metaphysics has been decided. The One as unifying
unity becomes normative for the subsequent determination
of Being.[2]

Therefore, it is because the One normatively decides on Being
that the latter is reduced to what is common, reduced to empty
generality. This is why it must also endure the metaphysical preeminence of beings.

Metaphysics can be defined as follows: the *enframing* of Being
by the One. Its most appropriate synthetic maxim comes from Leibniz.
The maxim ascertained the reciprocity of Being and the One as a
norm: "What is not truly *one* being is not truly a *being* either."[3]

The question from which I began speculating can now be
formulated as follows: Can the One be unsealed from Being?
Can the metaphysical enframing of Being by the One be severed
without in turn becoming involved in the Heideggerian idea of
destiny, or without entrusting thought to the unfounded promise
of a redemptory returning? For, with Heidegger himself, the
thinking of metaphysics as a history of Being is bound to an
announcement whose ultimate expression is that "only a god can
save us."

Can thought be saved without having to appeal to the proph-
ecy of a return of the gods? For that matter, has thought not always
saved itself, by which I mean: saved itself from the normative power
of the One?

In his *Introduction to Metaphysics,* Heidegger declares that
"on Earth an obscuring of the world comes forth."[4] He draws up a
list of the essential events of this obscuring: the flight of the gods,
destruction of the Earth, gregarization of Man, and preponderance
of the mediocre. All of these themes are coherent with determining
metaphysics as the exasperated normative power of the One.

If, through an original scission in its disposition, thought as
philosophy has forever marshaled the normative power of the One
simultaneously to seeking recourse against this power, that is, a
subtraction from this power, then the following ought to be said.
Just as an obscuring of the world forever comes forth, so also at the
same time does its enlightening. As such, the flight of the gods is
also the beneficial leave given to them by humankind; the destruc-

tion of the Earth is also its adjustment so as to be appropriate to
active thought; gregarization is also the egalitarian irruption of the
masses onto history's stage; and preponderance of the mediocre is
also the radiance and density of what Mallarmé used to call *action
restreinte,* "special action."[5]

Our problem then becomes figuring out how thought can designate from within itself the perennial effort to subtract Being from
the influence of the One? How can we come to terms with the fact
that, along with Parmenides, Democritus also existed and that with
him the dismissal of the One occurs through a dissemination and an
appeal to the void? How is the Heideggerian idea of destiny played
off against what are evidently exceptions to it? An example of such
an exception would be the magnificent figure of Lucretius. With
him, far from maintaining the appeal to the Open in distress, the
power of the poem attempts instead to subtract thought from any
returning of the gods and ascertain it in the steadfastness of the
multiple. Lucretius is the one who directly confronts thought to
subtraction from the One, which is none other than inconsistent
infinity, that is, what nothing can collect:

Such is the nature of the place, of the gigantic space:
Were it to slide, forever drawn away by time,
Lightning would never see distance reduced
The whole enormous reservoir of things is open
In all directions[6]

What has motivated me is to invent a contemporary version of
fidelity to what has never surrendered to the historical constraint of
onto-theology or to the enframing power of the One.

My initial decision was to contend that what can be thought
of Being per se is found in the radical manifold or a multiple that
is not under the power of the One. In *L'Être et l'événement* I called
the latter a "multiple without-One."[7]

Yet maintaining this principle involves highly complex requirements.

First and foremost, pure multiplicity or the manifold unfolding
the unlimited reserve of Being as a subtraction from the power of
the One, cannot be consistent on its own. In fact, we have to assume, as did Lucretius, that manifold-unfolding is not constrained

by the immanence of a limit. For it is only too obvious that such a constraint proves the power of the One as grounding the multiple itself.

The manifold as the exposure of Being to thought has to be posited then as not lying within the figure of consistent delimitation. Or rather ontology, if it exists, has to be the theory of inconsistent multiplicities as such. This means that what lends itself to the thought of ontology is a manifold without a predicate other than its own multiplicity. It has no concept other than itself, and nothing ensures its consistency.

More radically, a science of Being as really subtractive Being must prove from within itself the One's powerlessness. The without-One of the manifold cannot make do with a simple external refutation. Release from the One's grip is accomplished in the inconsistent composition of the manifold itself.

This point was grasped in its subsisting difficulty by Plato in the *Parmenides* when examining the consequences of the hypothesis that "One is not." This hypothesis is especially interesting when juxtaposed to Heidegger's determination of the distinctive character of metaphysics. What did Plato say? First of all, that if One is not, it follows that the immanent otherness of the multiple becomes an unending self-to-self differentiation. It is the astonishing formula: *ta alla etera estin,* which may be translated as: "the others are Others." The first other, marked by a lower case "o," contrasts with the Lacanian capital "O" distinguishing the second. The outcome of "the One is not" is "the other is Other" in terms of an absolutely pure manifold, a complete dissemination of itself. The motif of inconsistent multiplicity is found here.

Plato then goes on to show how this inconsistency dissolves the One down to the root of all of its supposed power—be it even the power to withdraw or inexist. All apparent exposure of the One immediately resolves it into an infinite manifold. Quoting from Plato: "To whoever thinks closely and sharply, every One appears as a limitless multiplicity, as soon as the One—since it *is* not—ends up lacking from it."[8]

What does Plato mean here if not that when subtracted from the metaphysical influence of the One, the multiple cannot be exposed to the thinkable as a multiple consisting of ones? It must be agreed that the multiple always and only consists of multiples. Every multiple is a manifold of multiples.

So long as a multiple (a being) *is not* a manifold of multiples, subtraction has to be maintained until the very end. One should not have to concede that such a multiple is the One, nor even that it consists of ones. That is the point at which it will inevitably be a multiple of nothing.

The subtractive also amounts to the following: instead of conceding that for want of the multiple there is the One, assert that for lack of the multiple there is nothing. This is how we end up running into Lucretius again. Lucretius did exclude the case of some such third property being ascribed to the One, somewhere between multiple compositions of atoms and the void:

> In addition to the void and bodies, there remains
> Among things no other nature
> That falls forever under our senses or that a mind
> Ends up discovering through reason.[9]

Moreover, this exclusion is what organizes Lucretius' critique of various unitary principled cosmologies, like that of Heraclites' Fire. Lucretius clearly sees that subtraction from the fear of the gods requires that, short of the multiple, there be nothing. Beyond the multiple there is still only the multiple.

Finally, a third consequence of a subtractive commitment is to exclude the following: that there could be a *definition* of the multiple. On this matter, Heideggerian discipline can help us: the strictly Socratic mode of carving out the Idea is performed by grasping a definition. The avenue of definition is opposed to the imperative of the poem precisely in that it lays out within language itself the normative power of the One. The Idea is to be thought in its being insofar as it is carved out, or frozen, by the dialectical resource of definition. Definition is the linguistic mode of ascertaining the pre-eminence of beings.

Were we to claim access to the multiple-exposure of Being by means of a definition, or by the dialectical route of successive delimitations, we would then be placed originally under the metaphysical power of the One.

Therefore, the definitional path is blocked for the thought of the multiple-without-One or the inconsistent manifold.

Ontology lies in the difficult passage whereby it has to expose the thinkable dimension of the pure multiple without ever being able to state the specific conditions a multiple affords. It is not even possible to make this negative duty explicit. For example, it simply cannot be said that thought is devoted to the multiple and to nothing but the intrinsic multiplicity of the multiple. Otherwise this thought would then already enter into what Heidegger calls the "process of limiting Being" by appealing to a delimiting norm. And the One would return.

It is not possible to define the multiple, or to make this absence of definition explicit. Actually, the thought of the pure multiple must be determined in such a way as to avoid mentioning the name "multiplicity" (or "manifold"). The name "multiplicity" should be used neither to say what it designates according to the One, nor to say, still according to the One, what it is powerless to designate.

Yet what is a thought that never defines what it thinks? That never exposes it as an object? A thought prohibiting itself from resorting to any name whatsoever of that thinkable, and in the very writing by which it is linked to the latter? Clearly, it is an *axiomatic* thought. An axiomatic thought seizes upon the disposition of undefined terms. It never encounters a definition of these terms or a practicable explication of what is not them. The primordial statements of such a thought expose the thinkable without making them thematic. Doubtless, the primitive term or terms are themselves inscribed, but they are not inscribed as a kind of naming in which a referent would have to be represented. Rather, this inscription points to the sense of a series of dispositions wherein the term lies only in the regulated game of its founding connections.

The most intimate requirement of a subtractive ontology is that its explicit presentation be in the form of an axiom, which prescribes without naming—but not as a dialectical definition.

It is from the standpoint of this requirement that the famous passage from the *Republic,* in which Plato opposes dialectics to mathematics, ought to be reinterpreted.

Let us read the summary Glaucon, one of Socrates' interlocutors, gives of his master's thought on this topic:

> Regarding being and the intelligible, the act of theorizing such as it is based on the science (*episteme*) of dialectics is clearer than the act based on what are called the sciences (*techne*). To be sure, those who theorize according to these

sciences, the first principles of which are hypotheses, are compelled to proceed discursively and not empirically. But as their intuiting rests upon hypotheses and fails to open access to first principles, they do not appear to you to possess the intellection of what they theorize. However, in light of first principles, this intellection stems from the intelligibility of beings. It seems to me that you call discursive (*dianoia*) procedures those used by geometers and their ilk; but not intellection in terms of the discursivity ascertained between (*metaxu*) opinion (*doxa*) and intellect (*nous*).[10]

It is perfectly clear that, for Plato, the trouble with mathematics is precisely the axiom. Why? Because the axiom remains outside of the thinkable. Geometers are compelled to proceed discursively because they do not enter into the normative power of the One, whose name is *first principle*. And this constraint is testimony to their exteriority in relation to the principle-based norm of the thinkable. In Plato's view, the axiom is laden with an obscure type of violence, owing to the fact that it does not appropriate the dialectical and definitional norm of the One. In the axiom and mathematics there is certainly thought, but not yet free thought. It is marshaled by the paradigm, the norm and the One.

On this issue, my conclusion is opposed to Plato's. What determines the axiom's value—the axiomatic disposition—is precisely its subtraction from the normative power of the One. Unlike Plato, I do not see the constraint it includes as being the sign of anything lacking in the unifying and grounding clarification he seeks. In it I see the necessity of the subtractive gesture itself. It is the movement by which thought is torn from everything that still keeps it connected to the common or to the general. Thought's own metaphysical temptation rests upon the latter at the cost of the inexplicit or of the impotency of naming. It is in this tearing away that I read the freedom of thought with respect to that to which it is destined to be constrained. It is something we could easily call its "metaphysical penchant."

Let us say that, in its dedication to the axiomatic disposition, ontology, or the thinking of the inconsistent pure multiple, cannot be guaranteed by any principle. Conversely, when treading back to the first principle, the multiple ceases to be exposed according to the immanence of its multiplicity alone.

There are five conditions for any ontology of the multiple to be conceived in its defection from the One's power. These conditions also stand for any ontology that is faithful to the struggle philosophy has waged against its own metaphysical tendency.

1. Ontology is the thought of the inconsistent manifold, that is, of what is reduced without an immanent unification to the sole predicate of its multiplicity.

2. The multiple is radically without-One in that it itself consists only of multiples. What there is, or the exposure to the thinkable of what there is under the sole requirement of the 'there is,' are multiples of multiples.

3. Granted that no immanent limit related to the One determines multiplicity as such, there is no first principle of finitude. The multiple can thus be considered infinite. Or even, infinity is another name of multiplicity as such. As no first principle binds infinity to the One, it ought to be tenable for there to be an infinite amount of infinites, an infinite dissemination of infinite multiplicities.

4. Given that a multiple can be considered as *not* being a multiple of multiples, we should withhold on reintroducing the One here. Instead, let us consider a multiple to be a multiple of nothing. And 'nothing' will be endowed with a consistency principle, though not anymore than multiples themselves will.

5. Actual ontological presentation is necessarily axiomatic.

At this point, enlightened by the Cantorian grounding of mathematics, we can assert ontology to be nothing other than mathematics itself. This has been the case ever since its Greek origin. However, mathematics has managed only with considerable difficulty and at the cost of toil and tiresome recasting to ensure the free play of its own conditions. Ever since its Greek inception, ontology has struggled within itself against the metaphysical temptation.

It can be said that with Cantor we move from "special ontology," which still links the multiple to the metaphysical theme of representing objects, numbers and figures, to "general ontology," which sets the free, thoughtful apprehending of multiplicity as such as the basis and destination of mathematics. It forever ceases to constrain the thinkable to the special dimension of "object."

Notice how post-Cantorian mathematics becomes somewhat equal to its conditions.

1. In Cantor's sense, the *set* has no other essence than to be a manifold. It has no external determination since nothing limits the way it seizes another thing. Nor does it have an internal determination, for that of which it is the multiple recollection is itself irrelevant.

2. In Zermelo and Fraenkel's stabilized elaboration, there is no other nondefined primitive term or value possible for the variables apart from sets. Hence, every element of a set is itself a set. This accomplishes the idea that every multiple is a multiple of multiples, with no reference to units of any kind.

3. Georg Cantor fully recognized not only the existence of infinite sets, but also the existence of infinitely many such sets. This infinity is itself absolutely open-ended. It is sealed only by the impossible and therefore real point that makes it inconsistent, namely, knowing the set of all sets cannot exist. In fact, this accomplishes Lucretius' a-cosmic philosophy.

4. There does exist a set of nothing, or a set that has no multiple as an element. This is the empty set. It is a pure mark from which all multiples of multiples are woven. The equivalence of Being and the Letter is thus achieved so long as there is subtraction from the normative power of the One. Let us muse over Lucretius' other powerful anticipation in the First Canto, verse 910ff:

A slight transposition suffices for atoms to create
Bodies both igneous and ligneous. It is like with words
When separating the letters somewhat,
We expressly distinguish igneous from ligneous.[11]

Were we now to refer to Lacan, it would be in this instance of the letter—an instance borne out by the mark of the void— that the thought unfolds, that is, the thought 'without-One,' or without metaphysics, of what lends itself to mathematical exposure as an immemorial figure of Being.

5. As the core of its presentation, Set Theory is nothing else than the theory's body of axioms. The 'set' is not a part of it, let alone is the definition of that word. The upshot of this is that the essence of the thought of the pure multiple does not require a dialectical principle. Furthermore, the freedom of thinking in harmony with Being is in the axiomatic decision—and *not* in the intuition of a norm.

The Cantorian presentation of sets was subsequently shown to be not so much a particular theory, than the very space of the mathematically thinkable. It is the famous "paradise" about which David Hilbert once spoke. Accordingly, a general retroactive move authorizes us to state the following. Ever since its Greek origin, Being per se has been insistently inscribed in the dispositions of pure mathematics. So it is from the very outset of philosophy that thought is subtracted from the normative power of the One. The striking incision of mathematics into philosophy from Plato to Husserl and Wittgenstein must be deciphered as a singular condition, that is, the condition exposing philosophy to the test of a way other than that of the subjugation of Being by the power of the One. From the outset and under its mathematical condition, philosophy is thus the scene of a disparate or split endeavor. It is true that philosophy exposes the category of truth to the unifying and metaphysical power of the One. And it is no less true that philosophy exposes this power to the subtractive defection of mathematics. This is why every singular philosophy is less of an actualization of its metaphysical destiny than it is an endeavor, under the mathematical condition, to be subtracted from it. The philosophical category of truth simultaneously results from a kind of normativity inherited from the Platonic gesture and from grasping the mathematical condition that ends up unraveling this norm. Besides, this is true with Plato himself. The progressive pluralization or blending of the supreme Forms in the *Sophist* or *Philebus,* as much as the act of bringing the theme of the One to an impasse in the *Parmenides,* both bear out the option between definition and axiom, principle and decision, and unification and dissemination as undecided and mobile.

In more general terms, if ontology, that is, what can be said of Being *qua* Being, is coextensive to mathematics, what then are philosophy's tasks?

Undoubtedly, the first task runs counter to its own latent vow, which is to humble itself before mathematics by acknowledging the latter as the thought of pure Being, of Being *qua* Being in its very act.

I stress that this is its *latent* vow because in its real becoming, philosophy has only had too great a tendency to claim mathematics does not accede to the status of genuine thought despite having had to examine it as admittedly necessary for its own existence. On this matter philosophy yielded to the sophistic injunction. This is partly responsible for the reduction of mathematics to the simple rank of computation or technology. It is a ruinous image for mathematics— one to which current opinion readily reduces it with the aristocratic complicity of mathematicians themselves. Mathematicians have willingly settled on believing that common folk understand nothing of their science.

It is philosophy's task to argue that mathematics *is a thought*. This is something philosophy has often tried to do, only to cancel out its task in the very same stroke.

2

MATHEMATICS IS A THOUGHT

There is nothing quite obvious about this statement. It has been asserted time and time again. First by Plato, who appended it with all kinds of reserves. It has also been negated time and time again: especially by Wittgenstein. The statement surely evades any attempts at proving its validity. Perhaps is it the dead end of mathematics itself, and therefore the Real of mathematics. But the Real is declared, instead of known.

The statement's obscurity is the outcome of what an "intentional" conception of thought seems to impose upon itself. According to this conception, all thought is the thought of an object that determines its essence and style. The intentional stance then goes on to posit mathematics as a thought exactly insofar as mathematical objects exist and philosophical investigation focuses on the nature and origin of these objects. Clearly, this type of presupposition is a problematic one. In what sense can mathematical idealities be declared to exist? And how can they exist in the generic form of object? Aristotle tackles this difficult problem throughout Book M of the *Metaphysics,* which he names the *mathematika,* that is, the mathematical things or presupposed correlates of mathematical science. In my opinion, Aristotle's solution is definitive in as much as the question of mathematics as a thought is dealt with from the angle of object or objectivity. His solution is inscribed between two limits.

1. On the one hand, being or existence can in no way be granted to mathematical objects in the sense that having this "being" is separate and comprises a preexisting and autonomous domain of objective donation. What Aristotle criticized here is a thesis usually attributed to Plato. It is a

fact that Aristotle's real descendants, namely, modern Anglo-American empiricists, call "Platonism" the supposed separate and supersensible existence of mathematical idealities. Counter to this supposition, they stress that mathematical objects are *constructed*. Aristotle would really be saying: the *mathematika* are in no way separate beings. Otherwise, there would be an original intelligible intuition of them to which nothing attests. They cannot be used to identify mathematics as a singular thought. For Aristotle, then, no ontological separation can guarantee epistemological separation, especially when related to the gap between physics, with its focus on the sensible, and mathematics, since: "it is manifestly impossible for mathematical things to have a separate existence from sensible beings."[1]

2. Symmetrically, it is just as impossible for mathematical objects to be immanent to the sensible. Aristotle deals with this point in Book B. The main argument is that the immanence of indivisible idealities to sensible bodies would entail the indivisibility of bodies. Or, that the immanence of immobile idealities would entail the immobility of sensible bodies, which is obviously refuted by experience. The incontestable core of this thesis is that all immanent mathematicity either infects the mathematical object with sensible predicates that are manifestly foreign to it, as with temporality or corruptibility; or it infects sensible bodies with intelligible predicates that are just as foreign to them, as with eternity or conceptual transparency.

With respect to experience, the mathematical object is neither separate nor inseparable. It is neither transcendent nor immanent. The truth is that strictly speaking it has no being. Or more precisely, the mathematical object exists nowhere *as an act*. As Aristotle would say, either the *mathematika* absolutely do not exist, or at any rate they do not exist absolutely. Mathematical objectivity can be deemed a pseudo-being, suspended between a pure separate act, whose supreme name is God, and sensible substances, or actually existing things. Mathematics is neither physics nor metaphysics.

But then what is it exactly? In fact, mathematics is a fictive activation (*activation fictive*), in which existence in act is lacking.

Mathematical objectivity exists potentially in the sensible. It remains there in the definitive latency of its act. As such, it is true that a person potentially grasps the arithmetic "One," or that a body potentially grasps one pure form or another. This is not to say that the arithmetic-One or the geometric sphere exists on their own, nor that they exist as such in a person or on a planet. The fact is thought can activate the One or the sphere from the experience of an organism or object. What does "activate" mean? Precisely this: to treat an existent as if it were an act when it only exists potentially. It means treating a being as a pseudo-being or taking something as separate that is not. This is Aristotle's own definition: the arithmetician and geometer obtain excellent results "by taking as separate what is not separate."[2]

The consequence of this fiction is that the norm of mathematics cannot be the true. The true is not accessible by a fiction. The norm of mathematics is the beautiful. For what 'fictively' separates the mathematician is above all relations of order, symmetries and transparent conceptual simplicities. Admittedly, Aristotle observes that "the highest forms of the beautiful are order, symmetry and the definite."[3] The upshot is that "the beautiful is the main object of mathematical proofs."[4]

Aristotle's definitive conclusion can be modernized. To this end, it suffices to ask oneself: what has the power to activate "potential being"? Or, what has the power to separate the inseparable? For moderns like us, it should be obvious that the answer to these questions is language itself. As Mallarmé once remarked in a fine citation, if I say "a flower," I separate it from every bouquet. If I say "let a sphere," I separate it from every spherical object. At that point, matheme and poem are indiscernible.

We can recap this doctrine as follows:

1. Mathematics is a pseudo-being's quasi-thought.

2. This pseudo-being is distributed in quasi-objects (for example, numbers and figures, as well as algebraic and topological structures, etc.)

3. These quasi-objects are not endowed with any kind of existence in act, as they are neither transcendent to the sensible nor immanent to it.

4. They are in fact linguistic creations fictively extracted from the latent or inactionable, or non-separable strata of real objects.

5. The norm managing the fiction of separation is the transparent beauty of the simple relations it constructs.

6. Mathematics is thus ultimately a rigorous esthetics. It tells us nothing of real-being, but it forges a fiction of intelligible consistency from the standpoint of the latter, whose rules are explicit.

And finally:

7. Taken as a thought, mathematics is not the thought of its own thought. Indeed, as set in its fiction, *it can only believe* in its thought. This is a point on which Lacan rightly insisted: the mathematician is first and foremost someone who believes in mathematics "hard as rock." The mathematician's spontaneous philosophy is Platonism because, as its act separates the inseparable, it draws the ideal spectacle of its result from this fictive activation. It is as though mathematical objects existed as acts. More profoundly, mathematical thought, as with any fiction, is an act. It can only be an act because there is nothing to contemplate within it. As Aristotle says in a very tight formula: with mathematics, (*e noésis energeia*), the intellect is in act. In mathematics, *the act lacking in objects ends up returning on the Subject's side.*

Caught in the act of fictive activation, which is none other than her own thought, a mathematician fails to recognize its structure. This is also why the esthetic dimension is dissimulated under a cognitive claim. The beautiful is the real cause of mathematical activity. But in mathematical discourse this cause is an absent one. It can only be spotted by its effect: "It is not grounds enough to claim that just because the mathematical sciences do not name the beautiful that they are not in fact dealing with it. For they show its effects and relations."[5] This is why it is up to the philosopher to name the real cause of the mathematical act. It is up to the philosopher to think mathematical thought according to its real destination.

In my view, the aforementioned conception of mathematics is still dominant today. It appears in four major symptoms:

(a) The critique of what is presupposed under the name of "Platonism" is more or less a consensus in all contemporary conceptions of mathematics. For much the same reason one can spot mathematicians who are spontaneous Platonists, or "naïve" ones.

(b) The constructive and linguistic character of mathematical entities or structures is almost universally accepted.

(c) Even if aesthetics is not summoned per se, many current themes are homogenous to it. As such, the absence of the category of truth; the tendency to relativism (there would be several different mathematics, and eventually it comes down to a matter of taste); last, the logical approach of mathematical architectures, which treats them as large forms whose construction protocol would be decisive and whose reference or very being is the determination in thought of *what* is thought and remains inassimilable. We are in line with Aristotle's orientation here. He explicitly recognizes a formal superiority to mathematics, which he calls a logical precedence, but he does so to better deny its substance or ontological precedence. As he says: "Substantial precedence is the sharing of beings who, insofar as separate, prevail through the faculty of separate existence."[6] The purely fictive separation of the mathematical object is thus, in terms of ontological dignity, inferior to the real separation of things. By contrast, the logical transparency of mathematics is esthetically superior to the separate substantiality of things. This is entirely reproduced today in the canonical intralinguistic distinction between the formal and empirical sciences.

(d) Nowadays, there is an incontestable supremacy of the constructivist, indeed intuitionistic vision, over the formal and unified vision of the ground, as well as on the self-evidence of classical logic. The great edifice undertaken by Nicolas Bourbaki was built according to a global esthetic that could be called "arborescent."[7] On the solid trunk of logic and a homogenous theory of sets, symmetrical branches of algebra and topology grew. They would end up crossing again in the higher reaches with the finest "concrete" structures, which in turn comprised a ramified lay out of foliage. Nowadays, one leaves off from already complex concretions instead. It is about folding and unfolding them according to their singularity, and of finding the principle of their deconstruction-reconstruction, without being concerned with an overall blueprint or with a pre-decided fundament. Axiomatics has been left behind to the benefit of a mobile grasping of complexities and surprising correlations. Deleuze's rhizomatics prevails over Descartes' tree. The heterogeneous lends itself to thought more than the homogenous. An intuitionistic or modal logic is more appropriate to this descriptive orientation than is the stiffness of the classical logic whereby the middle is excluded.

So the question is the following. Regarding mathematics as thought, are we fated to a linguistic version of Aristotelianism?

This is not my conviction. The injunction of contemporary mathematics seems to me to uphold Platonism. It strives to understand its real force, which has been completely ensconced by Aristotle's exegesis.

However, at this point, I shall not immediately embark upon what could be called a Platonist rectification. The question I should like to deal with and outline is more limited. Since it is all about ultimately getting to the thought of the thought of mathematics as a thought, it is relevant to point out the moments in which mathematics has apparently been convened to think itself, that is, to *say* what it is. As we know, these moments are conventionally termed "crises" or "foundation crises."

Such was the crisis of irrational numbers in so-called Pythagorean mathematics. Or the crisis linked to the "paradoxes" of Set Theory at the end of the nineteenth century, and the various limitation theorems in the formalism of the 1930s. There was also crisis with the anarchic handling of the infinitely small at the beginning of the eighteenth century, as well as crisis dealing with geometry upon the discovery of the undecidable nature of Euclid's postulate on parallel lines.

At the time, the question was often raised as to whether these crises were internal to mathematics or strictly philosophical. Imported into the debate among mathematicians were thought options linked to the existence of what Louis Althusser once called "the spontaneous philosophy of the scientists."[8] Althusser argued that there was no kind of crisis in the sciences. There were certainly discontinuities leading to hasty qualitative reshuffling, which did constitute moments of progress and creation. But in no way were they about dead-ends or crises. As these ruptures emerged, struggles in philosophical tendencies would ineluctably break out, affecting the respective scientific fields. The stakes in such struggles involved resetting the way philosophical currents help themselves to the sciences for their own ends and purposes. In the final analysis, the nature of these struggles is political.

Let us leave off from the following observation. There are singular moments when mathematics seems to be called upon, for its own purposes, to think its thought. What does this operation consist in? Everything is played out in a few statements that are the sticking points of mathematical thought, as if they were the signature of the impossible within its own field.

These statements come in three types:

One kind of statement can be a formal contradiction, drawn deductively from a set of presuppositions whose evidence and cohesion, however, appear to lie beyond doubt. Paradox is its sticking point. This is what happened to formal class theory in Frege's style, which stumbled on Bertrand Russell's paradox. The evidence held to be impossible here is of the kind that assigns to any property the set of terms possessing that property. There is nothing clearer than Frege's doctrine of the extension of a concept. Still, cases do tend to arise which are played out as real trials, affecting self-evidence with intrinsic inconsistency.

The second kind involves statements in which an established theory is diagonally crossed at one point by an exception or excess. Cases such as these compel a theory that had once been conceived as general to prove to be only local, indeed completely particular or "special" (*restreinte*). This is what happened to the demonstration stipulating that the diagonal of a square is incommensurable with its side, provided that "measurement" be understood to mean a rational number. The self-evidence of this assignation to all graspable relations of a pair of natural numbers once ensured the reciprocity between Being and Number for Pythagoreans. It was ruined demonstratively by a geometric relation exceeding any pair of natural numbers that could be assigned to it as a yardstick. Therefore the thought according to which Being's essential numericity obtains ought to be thought over again. This entails that mathematical thought as such ought to be thought over again.

Finally, the third case is when a previously unperceived statement is isolated as the very condition of results held to be certain, but when considered alone this statement seems unacceptable with respect to the shared norms of the constructions of mathematical thought. This was the case with the axiom of choice. The great late-nineteenth century French algebraists used to make implicit use of the axiom in their own demonstrations. But formally explicating the axiom appeared to totally exceed what they had accepted with respect to handling the infinite. They saw the axiom especially as illegitimately transgressing the constructive vision to which they subjected the operations of mathematical thought. The axiom of choice actually amounts to admitting an absolutely indeterminate infinite set whose existence is asserted albeit remaining linguistically indefinable. On the other hand, as a process, it is unconstructible.

It can be argued that mathematical thought returns unto itself under the constraint of a real sticking point or the inevitable emergence of an impossible point within its own field. This sticking point can be either of a paradoxical nature bringing about inconsistency, of a diagonal nature triggering excess, or of the nature of something disclosing a latent statement that brings forth the indefinite and unconstructible.

What then is the nature of mathematics as it twists unto itself beneath the injunction of an intrinsic sticking point? What surfaces concerns everything pertaining to an act or to decision making within the scope of mathematical thought. By the same token, a position has to be taken. For we stand actually as an act (*au pied de l'acte*), if I dare say, upon the very norm of the decision the act accomplishes.

At any rate, what is referred to in this obligation to decide is Being. Or it is about the mode according to which mathematics assumes what Parmenides himself said: "The Same is both thought and Being."[9]

Let us return to our examples again. For the Greeks, beneath the injunction of real incommensurability, thought was compelled to decide upon another way of knotting together Being and number, and the geometric and arithmetic. The name evoking this decision is Eudoxus. As for Russell's paradox, one had to accept a restriction on the powers of language to determine the pure multiple. The name evoking that decision is Ernst Zermelo. With the axiom of choice, thought was summoned to abruptly decide upon the indeterminate actual infinite—a decision that has divided mathematicians ever since.

In each case, it was about deciding on how, and according to what limits of an immanent disposition, mathematical thought is coextensive with Being. It had to do with how Being sustains the consistency of mathematical thought.

Therefore, when abutting against paradox and inconsistency, the diagonal and excess, or an indefinite condition, mathematics can be said to end up thinking that which, in its thought, relates to an ontological decision. What we are dealing with strictly speaking is an act. It is an act durably engaging the real of Being. The decision regarding the latter takes on the task of ascertaining its connections and configurations. Yet confronted in this way to its decisional dimension, mathematics can only become prey to the question of its norm. More accurately, it is prey to the question of the norm of what thought is able to sustain as an assertion of existence. Do we have to confer existence on numbers whose principle is to no longer consist of units? Must existence be conferred to nondenumerable actual infinite sets? What conditions can ensure a well-formed concept to accept an identifiable extension? How are the assertion of existence and the construction protocol linked? Can the existence of intelligible configurations be granted despite the impossibility of displaying even a single case of them? These questions will be settled according to an immanent norm that does not constitute thought, but orients it.

I call *an orientation in thought* that which regulates the assertions of existence in this thought. An orientation in thought is either what formally authorizes the inscription of an existential quantifier at the head of a formula, which lays out the properties a region of Being is assumed to have. Or it is what ontologically sets up the universe of the pure presentation of the thinkable.

An orientation in thought extends not only to foundational assertions or axioms, but also to proof protocols once the stakes are existential. Are we willing to grant, for example, that existence can be asserted based on the sole hypothesis that inexistence leads to a logical dead end? This is the spirit of indirect proofs, or reasoning *ad absurdum*. Granting its use pertains typically to the classical orientation in thought; not doing so pertains to the intuitionistic one. The decision has to do with what thought determines in and of itself as an access to what it declares as existing. The way towards existence sets up the discursive route to take.

To my mind, it is wrong to say that two different orientations prescribe two different mathematics, or two different thoughts. It is within a single thought that the orientations clash. Not one classical mathematician has ever cast doubt upon the recognizable mathematicity of intuitionistic mathematics. In either case, it is a question of a deep-rooted identity between thought and Being. But *existence*, which is both what thought declares and whose consistency is guaranteed by Being, is grasped according to different orientations. The fact is existence can be called that in respect of which decision and encounter, and act and discovery are indiscernible. Orientations in thought quite peculiarly take aim at the conditions of this indiscernibility.

It can thus be said that there are moments when mathematics, abutting on a statement that attests in a point of the impossible to come, turns against the decisions by which it is orientated. It then seizes upon its own thought, doing so no longer according to its demonstrative unity, but according to the immanent diversity of orientations in thought. Mathematics thinks its unity as *exposed* from within to the multiplicity of thought orientations. A "crisis" in mathematics arises when it is compelled to think its thought *as the immanent multiplicity of its own unity*.

I believe that it is at this point, and only at this point, that mathematics—that is, ontology—functions as a condition of philosophy. Let us put it this way: mathematics relates to its own thought according to its orientation. It is up to philosophy to pursue this gesture by way of a general theory of orientations in thought. That all thought can only think its unity by exposure to the multiplicity orientating it is a notion mathematics cannot accomplish alone, albeit this is what it manifests most exemplarily. The complete relation of mathematical thought to its own thought assumes that philosophy, under the condition of mathematics, deal with the question: what is an orientation in thought? What is it that imposes the identity of Being and thought to be carried out according to an immanent multiplicity of orientations? Why must one always decide upon what exists? The whole point is that existence is in no way an initial donation. Existence is precisely Being itself in as much as thought decides it. And that decision orients thought essentially.

One ought to have a theory of thought orientations at one's disposal, and use it as the real territory of what can activate the thought of mathematics as a thought. For my own purposes, I recommended a summary dealing of this point in *L'Être et l'événement,* though I cannot return to its technical substructure. Let it suffice for our purposes here to recall how three major orientations were put forward in that book. They can simultaneously be identified in moments of mathematical crisis and just as well in times of conceptual reshuffling within philosophy itself. These orientations are the "constructivist," transcendent and generic orientations.

The first sets forth the norm of existence by means of explicit constructions. It ends up subordinating existential judgment to finite and controllable linguistic protocols. Let us say any kind of existence is underpinned by an algorithm allowing a case that it is the matter of to be effectively reached.

The second or transcendent orientation works as a norm for existence by allowing what we shall coin a "super-existence." This point has at its disposal a kind of hierarchical sealing off from its own end, as it were, that is, of the universe of everything that exists. This time around, let us say every existence is furrowed in a totality that assigns it to a place.

The third orientation posits existence as having no norms, save for discursive consistency. It lends privilege to indefinite zones, multiples subtracted from any predicative gathering of thoughts, points of excess and subtractive donations. Say all existence is caught in a wandering that works diagonally against the diverse assemblages expected to surprise it.

It is quite clear that these orientations are—metaphorically—of a political nature. Positing that existence must show itself according to a constructive algorithm, that it is predisposed in a Whole, or that it is a diagonal singularity: all of these orient thought according to a repeatedly particular meaning of what is. And consideration of the 'what is' is here based on the decision to attribute existence. Either 'what is' is that of which there is a case, or 'what is' is a place in a Whole; or 'what is' is what is subtracted from what is a Whole. As for the claim I made above, we could transpose what I am speaking of in terms of a politics of empirical particularities, a politics of transcendent totalities and a politics of subtracted singularities, respectively. In a nutshell, they are embodied, respectively, as parliamentary democracies, Stalin, and as something groping forward to declare itself, namely, generic politics. The latter suggests a politics of existence subtracted from the State, or of what exists only insofar as it is incalculable.

What is otherwise wonderful is how these three orientations can be read mathematically by sticking to Set Theory. Gödel's doctrine of constructible sets gives a solid base to the first orientation; the theory of large cardinals provides one for the second orientation; and the theory of generic sets lends itself to the third.

Still, many other more recent examples show how every mathematical breakthrough ends up exposing these three orientations in the contingent unity of its movement. Every real movement is forthwith the presentation of the three orientations to the point of convening thought to it. Every real movement confronts the formal triplicity of these decisions regarding existence.

What we must bear in mind about this latter point, which is of great help in any concrete situation, can be summed up as follows. No serious quarrel between thought-apparatuses manages to oppose the interpretations of a kind of existence everyone recognizes. In fact, the opposite is closer to the truth. It is about existence itself that agreement fails to occur, as this is what had been decided upon in the first place. Every thought is polemical. It is no mere matter of conflicting interpretations. It is about conflicts in existential judgments. This is why no real conflict in thought reaches a full resolution. Consensus is the enemy of thought, for it claims we share existence. In the most intimate dimension of thought, however, existence is precisely what is *not* shared.

Mathematics has the virtue of not presenting any interpretations. The Real does not show itself through mathematics as if upon a relief of disparate interpretations. In mathematics, the Real is shown to be deprived of sense. It follows that when mathematics turns back upon its own thought, it bears existential conflicts. This ought to give us food for thought regarding the idea that every grasping of Being, as related to existence, presupposes a decision that decisively orients thought without any guarantees or arbitration.

Lautréamont's praise of "severe mathematics" (*mathématiques sévères*) is befitting.[10] What is severe is not so much its formalism or demonstrative entailment. Rather, it is the laying bare of a maxim of thought that can be formulated thus: it is when you decide upon what exists that you bind your thought to Being. That is precisely when, unconscious of it all, you are under the imperative of an orientation.

3

THE EVENT AS TRANS-BEING

Let us assume mathematics to be the thought of Being *qua* Being. Let us suppose that the latter comes into its own thought when the existential decisions prescribed by an orientation are at stake. What then can philosophy's own field be deemed to be?

We have already seen why it is up to philosophy to identify the ontological vocation of mathematics. Save for the rare moments of "crisis," mathematics thinks Being per se. Yet at those moments mathematics is not a thinking of the *thought* that it is. Admittedly, to be historically able to unfold as the thought of Being, and owing to how difficult it is to uproot it from the metaphysical power of the One which is encountered there, it could just as well be said there is no alternative but to identify mathematics as something completely different to ontology. It is up to philosophy to state and legitimize the following equation: mathematics=ontology. That said, philosophy manages on its own to be released from what is apparently its highest responsibility by asserting that it is simply not in charge of thinking Being *qua* Being. In fact, the movement by which philosophy is purified does not lie in its jurisdiction. It is all about identifying what its own conditions are, and this is a pattern scanning philosophy's entire history. Philosophy has been released from, or even relieved of, physics, cosmology, and politics, as well as many other things. It is also important for it to be released from ontology per se. But this task is complex largely owing to the fact that it entails that real mathematics be crossed reflexively—and not epistemologically. For example, in *L'Être et l'événement,* I simultaneously sought to:

- Examine the ontological efficiency of the axioms of Set Theory through the successive categories of difference, void, excess, infinite, Nature, decision, truth, and Subject;

- Show how and why ontological thought is accomplished without having to be identified as such;

- Examine, in the non-unified vision I propose of philosophy's destiny, the philosophical connections of axiomatic interpretations: Plato's *Parmenides* on the One and difference, Aristotle on the void, Spinoza on excess, Hegel on the infinite, Pascal on decision, and Rousseau on the being of truths, and so forth.

The way I see it is that the work to do here is still wide open. As Albert Lautman's works especially demonstrated already in the 1930s, every significant and innovative fragment of real mathematics can and ought to prompt its ontological identification in terms of a living condition.[1] I tried my hand at this myself recently with respect to the status of the concept of number in the renewed version John Horton Conway introduced and—more on this anon— with respect to the theory of categories and *toposes*.[2]

On the other hand, a vast question opens up regarding what is subtracted from ontological determination. This is the question confronting what is not Being *qua* Being. For the subtractive law is implacable: if real ontology is set up as mathematics by evading the norm of the One, unless this norm is reestablished globally there also ought to be a point wherein the ontological, hence mathematical field, is de-totalized or remains at a dead end. I have named this point the *"event."* While philosophy is all about identifying what real ontology is in an endlessly reviewed process, it is also the general theory of the event—and it is no doubt the special theory, too. In other words, it is the theory of what is subtracted from ontological subtraction. Philosophy is the theory of what is strictly impossible for mathematics.

Inasmuch as mathematics ensures Being as such in thought, the theory of event aims at determining a trans-being.

The question then becomes the following: granted that the event is that about which all is not mathematicized, do we have to conclude that the multiple is intrinsically heterogeneous? By taking the event as Being's breaking point—something I call the "structure of trans-being"—it does not exempt us from thinking the being of the event itself. The being of trans-being. Does this being of the event require a theory of the multiple which is heterogeneous to the theory explaining Being, that is, Being *qua* Being? I see Deleuze's position as hinting at this. Thinking the event fold originally demands a double-edged manifold theory, one that pursues Bergson's legacy. According to this view, extensive and numerical manifolds must be distinguished from intensive or qualitative multiplicities. An event is always the gap between two heterogeneous multiplicities. What occurs ends up making a fold, as it were, between the extensive spreading out and the intensive continuum.

As for myself, I contend the contrary, namely, that multiplicity is axiomatically homogenous. That's when I have to explain the being of the event both as a breach of the law according to which manifolds spread out as well as something homogenous to this law. This takes place through a defection of axioms: an event is nothing but a set, a manifold. But its emergence and supplementation subtract one of the axioms of the manifold, namely, the axiom of foundation.

Taken literally, this means an event is strictly an un-founded manifold. This defection is in fact a pure chance supplement to the manifold-situation for which it is an event.

This said, the general question of the status of the event in relation to the ontology of the manifold, over which Gilles Deleuze and I argued at length, and on how not to reintroduce the power of the One to such an extent that the manifold law ends up failing, is in my opinion the main question of any contemporary philosophy. Besides, this was all pre-constituted with Heidegger in the slide from *Sein* to *Ereignis,* from Being to his conception of the Event. We might also shift registers and have an altogether opposite glance at the question. Lacan invested it in the thought of the psychoanalytic act. He saw it as an eclipse of truth between presupposed knowledge and transmissible knowledge, or between interpretation and the matheme. This is also a decisive problem for Nietzsche: if the task is to break the history of the world in two, what is the thinkable principle of such a break in the absolute affirmation of life? It also happens to be Wittgenstein's central problem: how does the act open up onto a silent access to the "mystical element," that is, to ethics and aesthetics, if sense is always captive of a proposition?

In each of the aforementioned cases, the latent matrix of the problem is the following: if by *"philosophy"* one must understand both the One's jurisdiction and the conditioned subtraction from this jurisdiction, how can philosophy seize what is happening?

Namely, seize what is happening in thought? Philosophy will always be split between recognizing the event as the One's supernumerary coming, and the thought of its being as a simple extension of the manifold. Is truth what comes to Being or what unfolds Being? We remain divided. The whole point is to contend, for as long as possible and under the most innovative conditions for philosophy, the notion that truth itself is but a multiplicity: in the two senses of its coming (a truth makes a typical multiple or generic singularity befall) and of its being (there is not *the* Truth, there are only disparate and untotalizable truths that cannot be totalized).

A radical gesture is precisely required here. Besides, this is how modern philosophy is recognized: by subtracting the examination of truths from the simple form of judgment. What this has always meant is to decide upon a single ontology of manifolds. This is all about remaining faithful to Lucretius. It is all about telling ourselves every instant is one in which

> From all over an infinite space opens
> When atoms, innumerable and boundless,
> Flutter about in eternal movement.[3]

Despite his Stoic inflections, was Deleuze not himself faithful to Lucretius? I would like to return to what made Deleuze choose against ontological mathematicity in the end, and what made him choose the word "life" as Being's main name.

4

DELEUZE'S VITALIST ONTOLOGY

Deleuzian ontology issues a major injunction. It ensures that Being does not fold to any category, to any set disposition of its immanent partition. Being is univocal, provided that beings or existents are never distributed and are classified according to equivocal analogies.

For example, let us ask what is "sexuated," or sexual, being. It is impossible to construct an intuition of this if you base it upon the identification of man or the masculine. It is not easier if you leave off from feminine being, be it as exception or breach, or the supposed precedence of *feminitude*. What we need is to get to the inflection point at which the territoriality of man's becoming-woman and woman's masculine terroriality intertwine in a bifurcating topology. Man is thinkable only as an actualization of his feminine virtuality. Man is thinkable only at the point at which he is unassignable to masculinity, because his feminine virtuality is itself a line of flight out of masculine territoriality. As such, we think the sexual being when we lie in the indiscernibility between a movement of feminization and a suspension of masculinization exchanging their respective energies in the indiscernible.

It might just as well be said that sexual being thought according to its being, that is, according to modal activation, is not itself sexuated. It is not even sexual—if by 'sexual' we understand a repertory of properties. Whatever this repertory might be, and even were we to infinitely complicate it, the sexual being can only be intuited within that unattributable and indiscernible space in which all properties undergo constant metamorphosis.

That Being has no properties is an old thesis. But Deleuze's renewal of it consists of having Being be the active neutralization of properties through the unseparated virtualization of their effective separation.

That Being is not a property, that it is an "impropery," is also an old thesis. This is exactly what Plato means when asserting that the Good, which is the name of Being, is not an Idea. For every Idea is the actual-being of a property. The Good refers to no property as it is that from which any kind of property and any idea comes to the power of the partition it institutes.

Deleuze transforms the theme of Being's lack of properties, or its impropery. For he thinks that with Plato the Good's trans-ideal impropery still remains a property, namely, the transcendent property of the improper as such. How can Being be thought as impropriety without assigning a sort of transcendent super-property to it? How can we avoid seeing the improper ultimately turn into what is Being per se? The path he takes up is the one called univocity, or immanence. These are identical. One day Deleuze wrote to me in capital letters: "immanence=univocity." What is this all about? Being's impropery, or rather impropriety, is nothing but the defection of properties through their virtualization. By contrast, the properties of beings or existents are nothing but the terminal simulacrum of their actualization. Being is the "depropriation" of what belongs to property as such, that is, what is *proper* to it. It is also the appropriation of its own impropriety. This means that it is the movement of two movements. Rather, it is the neutral movement of the Whole, such that in it the partitioning of beings befalls according to the undividable, or indiscernible, of the movement disjoining them.

This is the fundamental reason for which Being deserves the name *life*. This is a real question. Why must Being, as univocity or immanence, be called "life?" Why is Being as potential (*puissance*) "the powerful inorganic *life* surrounding the world?" In philosophy, how to name Being always involves a crucial decision. It sums up thought. Even the name "Being," if chosen as the name of Being, involves a decision that is in no way tautological. This can be seen with Heidegger. The name given to Being induces a declension of names. So with Heidegger, the meanings of *Sein* and *Dasein* get caught in the movement of the so-called turn, which ends up also engulfing *Ereignis*. In my own account, this movement has the disjunctive series shift from the multiple to the void, and from the void to the infinite, and finally from the infinite to the event.

What is it with Deleuze that pins down the thought of Being to its Nietzschean name, life? The answer to this question is that

Being must be evaluated as potential, but as an impersonal or neutral potential.

It is potential because it is rigorously coextensive to the actualization of the virtual and to the virtualization of the actual, or indeed to the impropriation of the proper, and to the "propriation" of the improper. It is coextensive to the de-linking disjunction of multiple existents and to Relation, which defines the Whole. In this 'and,' in this conjunction, the moving gap must be thought as the movement of Being itself. The latter is neither a virtualization nor an actualization, but the indiscernible medium of the two, that is, the movement of two movements. This is a mobile eternity in which two diverging time sequences are bound.

This is the reason for which Being is neutral, too. Its power is to metamorphose what is presented as a categorical partition into the "eternal return of the same." It is to bring about what is assertively subtracted from the disjunctions it ceaselessly carries out. Being is a modalization by the middle of what appears to be distributed. As such, it does not allow itself to be thought in any distribution.

Here we have the deep Deleuzian sense of Nietzsche's statement "beyond good and evil." Good and evil are the moral or genealogical projections of any categorical partition whatsoever. Deleuze can be read as saying: beyond the One and the Multiple, beyond identity and difference, and beyond time and eternity. Especially: beyond the true and false. Yet "beyond" clearly does not mean a synthesis or a third transcendent term. "Beyond" means in the middle, wherein Being is what activates the essential falsehood of the true and virtualizes the truth of the false through the rhizomatic-network shifter between virtualization and actualization. Being is what prompts secret goodness to come forth. It brings about the infernal goodness of evil. It is what unfolds the evil spell of the good.

It is still poor and inaccurate to say that Being's neutrality is to be identified neither with evil nor with good, and neither with the false nor the true. This "neither ... nor ... " lacks the "and ... and ... " of metamorphosis. For Being is the becoming-false of the true, the becoming-true of the false. This is why it is neutral in terms of Being, as well as true and false.

Yet even the "and ... and ... " is still too poor, still too categorical according to this view.

As everyone knows, Deleuze despised logic. The linguistic and logicist turn of philosophy at the beginning of the twentieth century was a matter of considerable grief for him. The mortification of Melville and Whitehead's powerful Anglo-American world by the ruminations of analytic philosophy was a consternating spectacle for Deleuze.

Ever since Aristotle, logic has been but the numbering of categories and the triumph of property against impropriety. We have to dig another logic out from Deleuzian univocity. This is a kind of a logic in which the way categories are distributed according to the usual connections no longer satisfies us. The "and . . . and . . . ," or the "either . . . or . . . ," the "neither . . . nor . . .": all of these relations have extenuated and dilapidated Being's powerful neutrality. Instead, we ought to be thinking of a shifting superposition of the *and,* the *or* and the *nor.* This would allow us to then say that Being is neutral. It is neutral insofar as every conjunction is a disjunction, and every negation an affirmation.

Deleuze called this neutrality "and-or-nor" connector, the *disjunctive synthesis.* It must be said that Being as a neutral power deserves the name "life" because it is, in terms of a relation, the "and-or-nor," on the disjunctive synthesis itself. It is also the conjunctive analysis, the "or-and-nor." In fact, life specifies and individuates, it separates and unbinds. It incorporates, virtualizes and joins together, too. Life is the name of being-neutral according to its divergent logic, according to the "and-or-nor." It is a kind of creative neutrality holding ground in the middle of disjunctive synthesis and conjunctive analysis.

This is why Deleuze was the one to carry out the deepest thought on Nietzsche's capital idea. Nietzsche emphasized that life produces value gaps. Life is an evaluating power and is also active divergence. In and of itself, life cannot be evaluated, it is neutral. The value of life, says Nietzsche, cannot be evaluated. This is tantamount to asserting there is no life of life. It is only from the standpoint of life that some such existent can be evaluated. This is what univocity means: there is no being of Being. If the word 'life' as a proper name suits this univocity, then it evident that there cannot be the life of life, as it were. There is only life's movement, which is itself thinkable as an in-between of the movements of actualization and virtualization. This is why the power of Being, which is Being itself, is neutral, impersonal, unattributable, and indiscernible. The name "life" fits when assembling these "im-properties."

The short book I wrote on Deleuze had barely been published when the reproaches against me began.[1] It was as if my claim that Deleuze's philosophy involved an ascetic conception of thought, that it went against spontaneity and demanded a firm break with the ego's injunctions, was a paradox that could not be defended and was even lighthearted.

Let us ask then: what is the thought that can be equal to the task of Being's neutrality in line with the way intuitions are constructed of it? How can one reach the shifters and gaps in movement, the points that become impersonal, unattributable and indiscernible? How does one dissolve the closed pretensions of our actual-being in that big complete circuit of the virtual?

Deleuze is at least as consistent as Nietzsche. Admittedly, Nietzsche knew that everything must be asserted, and that the Dionysian Noon leaves no parcel of the Earth outside of its own thinking activation. For Nietzsche, once all figures of force are seized within the kernel of power, which reaffirms Dionysus' coming, they can be integrated to him. Thereupon Dionysus dismembers and recomposes himself in the laugh from which the gods died. Nietzsche knows that the name "life" names Being's complete equality. Deleuze asserts with him that Being is equality itself. How could noncategorical neutrality be unequal? Nonetheless, Nietzsche ended up opting for aristocratism in thought and vouching for the supereminence of the strong. Surely, this might seem paradoxical. However, who or what is strong? Strong is the one who completely asserts Being's equality. Weak is the one standing unequally in this equality, mutilating and abstracting the joyous neutrality of life. Thus conceived, there is nothing self-evident about force itself. Force is concentration and effort. It is the unsheathing privation of all categories under which we construct the opaque shelter of our actuality, individuality and ego. "Sobriety, sobriety," they say in *A Thousand Plateaus*.[2] Sobriety because the spontaneous opulence, the derisory trust in what we are categorizes us as an impoverished and assigned region of Being. So, yes, this *is* asceticism and stoicism, because to be able to think we have to give ourselves the wherewithal by which to go beyond these limits, to go to the end of what we are able to do. Asceticism is what is at stake here, because life constitutes us and judges us "according to a hierarchy that considers things and beings from the point of view of power (*puissance*)."[3] Being worthy of inorganic life implies not attending too much to the satisfaction of our organs. The nomad is the one who knows not to drink when thirsty, strives ahead under a blazing sun when wanting only to sleep, and sleeps alone on the desert sand when dreaming of hugs and rugs. Nomadic thought blends with the neutrality of life and with metamorphosis by an enduring exercise during which we take leave from what we are.

Nietzsche's "become who you are" must be understood as you are only what you become. Getting to where the impersonal force of the outside activates this becoming engages us to treat ourselves as disjunctive syntheses and conjunctive analyses, that is, to separate and dissolve ourselves. This is how we can shed some light on the way "great health" is retrieved from illness. Health becomes an affirmation and metamorphosis, instead of a state and satisfaction. Here we have the hero of flexible speech through whom indiscernible life speaks. Or we have Beckett's hero: the exhausted one sawed in half, a head streaming with tears and planted in the sawdust of a jar. And you refuse to say that thought, life-thought, is asceticism?

There is something truly and terribly painful in Deleuze's thinking. It is the antidialectical condition of joy. It is a lessening of self so that Being might decline in its unique clamor by means of your mouth and hands.

The name of Being is Life. But it is so for who does not take life as a gift, treasure, or survival, but as a thought returning to where every category breaks down. All life is naked. All life is denuded, abandoning its garments, codes, and organs. Not that we are headed for the nihilist black hole. Quite the contrary, we stand at the point where actualization and virtualization switch places, so as to be a creator. This is what Deleuze calls a "purified automaton," an increasingly porous surface to Being's impersonal modalization.[4]

Where does the difficulty lie, then? Like with Nietzsche, I should say it is in Deleuze's sign theory, in "what makes a sign." It is what makes a sign for the impersonal in the personal, for the virtual in the actual, for the nomad in the sedentary, for eternal return in chance and for memory in matter. Synthetically, it is what makes a sign for the Open in the Closed.

Would I be faithful to Deleuze were I not to voice my reticence and resistance here? The fact is I am convinced nothing makes a sign. By maintaining its marks, even in minimizing it to the fullest, even as a differential infinite, Deleuze still concedes too much to a hermeneutics of the visible.

Nietzsche's sign theory is well known to be circular. This explains why Zarathustra must identify himself as his own precursor. He is the rooster in the street whose song announces his own coming. What makes a sign for the overman is the overman itself. The overman is only the sign in man of the overman's coming. There cannot be a distinction made between the event and its announcement. Zarathustra is Zarathustra's sign. Nietzsche's madness is to get to a specific point of indiscernibility at which breaking the world in two according to its announcement implies that one must also break oneself. That is the option once the sole sign of his "great politics," wherein the world is broken, is this meager singularity that under Nietzsche's name, a Nietzsche wandering the streets of Turin alone and incognito, declares its imminence.

Like Nietzsche, Deleuze has to tag the actual, closed and disjointed existents simultaneously with their co-belonging to the great virtual totality. Just as they also have to cancel the mark so that Being's neutrality is not found to be distributed in categories. The Closed must bear the sign for the Open. It must make a sign in and of itself for the Open. Otherwise how would one go about shedding light on what we think? How would we understand that we are obliged to "dis-close" our actuality? The sign of the open or of totality is that no closure is complete. As Deleuze says, "the set is always held open somewhere as if by the finest thread that connects it to the rest of the universe."[5] As fine as the thread might be, it is still Ariadne's thread. As such, it concentrates Deleuze's ontological optimism. As disjointed and closed as actual existents may be, a fine mark in them guides thought toward the total life to which they are disposed. Without the latter we would not be able to think the Closed according to the Open or from its virtuality. For nothing ever starts absolutely.

Yet at the same, there can be no signs. Nothing alone can be a sign. Otherwise Being would no longer be univocal. Being would be taken according to various senses, such as Being without being or Being *qua* Being, and the sense of Being according to the sign of Being. Which is why when Deleuze speaks of objects, he has to simultaneously affirm that they have a real part and a virtual part, and that these two parts are indiscernible. The upshot is that the virtual part of the object, which is precisely its opening and what makes a sign toward totality, is not really a sign after all. For its sign function cannot be discerned from that in relation to which it makes a sign. In truth of fact, the opening point of closed sets is even less than a fine thread. It is a component that is both entirely caught in the closure and yet completely open. Thought is unable to separate these two holds. Consequently, thought is unable to isolate the sign.

The univocity postulate conditioning the idea of the name of Being is life. To keep this postulate, Deleuze and Nietzsche have to posit that everything is like a sign of itself, in an obscure sense. Not of itself in terms of itself, but of itself in terms of a provisional simulacrum or precarious modality of the power of the Whole.

Yet if a thing is a sign of itself, and its sign dimension is indiscernible from its being, it makes no difference to say that everything is life, and that everything is a sign.

The name of Being is life if it is thought from the angle of the univocal power of sense. Being's name will be Relation if it is thought from the angle of the equivocal universal distribution of signs.

Existents themselves will be entirely disjoint and without relation if they refer to Being as inorganic life. They will be entirely connected and consonant if they refer to Being as Relation.

I believe the equivocal is reinstalled this way at the very heart of Being. The categorical distribution is expelled from large macroscopic classifications like the sensible and the intelligible. Yet perhaps it shows up again in the microscopic when the indiscernibility of the components of the existent shifts it equivocally toward the disjunctive synthesis of life or toward the conjunctive analysis of Relation.

This can also be said otherwise. Deleuze sets up an immense, virtuosic, and ramified phenomenological apparatus so as to be able to write the ontological equation as Being=event. Yet at the minutest point of what this apparatus is able to seize, what appears is precisely that the being of this Being is *never* the event. The upshot is that Being remains equivocal.

This is how, by seeking out to learn from this genius, I reached the conviction that the pure multiple, the generic form of Being, never itself greets the event as a virtual component. On the contrary, the event befalls unto it through a rare and incalculable supplementation.

To achieve this, I had to sacrifice the Whole, sacrifice life and sacrifice the great cosmic animal, whose surface was enchanted by Deleuze. Thought's general topology is no longer "carnal or vital," as he used to declare. It is caught in the crossed grid of strict mathematics, as Lautréamont used to say, and the stellar poem, as Mallarmé would have said.

In the end, the two great dice throwers of the late nineteenth century, Nietzsche and Mallarmé, individually chose their own. Still, the great philosophical passion of the game is universally shared. Indeed, that is what it is all about. Deleuze said it once and for all: thinking is throwing dice.

The only thing we ask is how a player of thought, a thrower of dice like Deleuze, can so insistently claim filiation with Spinoza, to the point of making him the "Christ of philosophy?"[6] Where does he make room in Substance's immanent necessity for chance and the game? No doubt, like many interpreters, Deleuze neglects the function mathematics holds in Spinoza's ontology. This is where we come full circle to our initial idea. It is not an overstatement to say that for Spinoza mathematics only thinks Being, for mathematics alone consists completely of adequate ideas. This is what I sought to ascertain by placing Spinoza at a divide—yet another one—between Deleuze and myself. Not that, for me, Spinoza should be philosophy's Antichrist! Spinoza is much rather philosophy's excessive Guardian.

5

SPINOZA'S CLOSED ONTOLOGY

When a thought-proposal on Being is presented from outside mathematics as originally philosophical, it deals with the general nature of the "there is." Thereupon three primordial operations are necessarily summoned.

First, the norm or norms of the "there is" have to be constructed and legitimated, as we saw with my "pure multiple" or Deleuze's "life." These names are always grasped more or less explicitly in an option having to do with the type of joint, or disjoining, between the One and the Many.

Next comes the task of unfolding the relation or relations from which evaluating the consistency of the "there is" is undertaken.

Last is the complex body of any philosophy of Being, which is considered here as implicit mathematics. What I call its "hold," or wedge, has to be laid out. These are the formal, intelligible relations of what is assumed or tolerated by the names of the "there is."

Let us give two typical and contrasting examples of names. The first one is poetico-philosophical in nature, and the other purely mathematical:

- In Lucretius' aforementioned undertaking, the "there is" is assumed under two names: void and atoms. The relations are shock and hitching. What ensures their hold on the nominal components of the "there is" is an unassignable event, that is the *clinamen*. The *clinamen* or swerve is that by which the indifference of the atom's trajectory befalls onto relations against a background of the void, and ends up constituting a world.

- In the mathematical theory of sets, which I have argued accomplishes mathematics as a thought on manifold-Being, the "there is" is assumed in the name of the void or empty set alone. The only relation is membership. The hold of the relation on the "there is" is guaranteed by the efficient forms of the relation, which are codified in axioms, that is, the theory's operative axioms. This hold draws a "universe" from the void alone: the cumulative transfinite hierarchy of sets.

Moreover, there are perhaps only two models of the hold and therefore of the thought-operation through which the names of Being comply with the relation lending them consistency: the event model, which is Lucretius,' as well as the axiomatic model.

Spinoza excluded the event by prohibiting excess, chance, and the Subject. As a result he opted for the axiomatic figure absolutely. From this point of view, the *more geometrico,* the geometrical manner, is crucial. This is not a form of thought, but how the writing of what amounts to an original decision in thought is traced.

A scholarly inspection of the *Ethics* brings to bear a strong sense of simplicity. The "there is" is indexed to a single name, namely, the absolutely infinite Substance or God. The only relation admitted is that of causality. The relation's hold on the name is of the order of the immanent effectuation of the "there is" itself. For as we know, "God's power is identical to His essence."[1] Not only does this mean that "God is the immanent cause of all things,"[2] but that His identity is thus insofar as it is thought through the hold of the causal relation.

Here we have an entirely affirmative, immanent, and intrinsic proposition on Being. As constitutive of Lucretius' ontology for example (there are the void *and* the atoms), the concept of difference would be absolutely subordinate, indeed nominal. It would stand as a question of expression that in no way alters the determination of the "there is" as lying under the sign of the One. Among some hundred different passages, it is worth citing this one: "A mode of extension and the idea of that mode are one and the same thing, but expressed in two ways (*duobus modis expressa*)."[3]

The simplicity is obvious but apparent. In fact, we would have to show:

- First, that there is a multiple and complex intertwining of what allows the naming of the "there is," and that the self-evidence of difference is constantly required in this intertwining;

- Second, that there is not only a single fundamental relation, causality, but at least three, which apart from causality I call "coupling" and "inclusion";

- Third, that under the "there is," an altogether exceptional kind of singularity is drawn out from the depths. Its formal characteristics are those of a Subject, whose Spinozian name is *intellectus*. Following Bernard Pautrat's persuasive suggestion, I translate this *intellectus* by "intellect"—maintaining the term in English, as I did in French.[4] We reach the heart of Spinozian ontology when understanding how this intellect requires propositions on Being which are *heterogeneous to the explicit propositions*.

In the *Ethics,* the naming of the "there is" is God, as previously mentioned. But the construction of this name—what Spinoza calls its "definition"–is extremely complex.

God is *"ens absolute infinitum,"* an "absolutely infinite existent." Note that the requisitioning of the indeterminate term *ens,* "existent," as what names a virtual "there is," is preunderstood as referring to an ontological stratum, if not to an even deeper or at least more extensive one. "Infinite" is obviously the big question, for here we find the determinant of the undetermined, and practically the "there is" of the "there is." The important thing is that the absolute character of divine infinity is not qualitative or itself indeterminate. It refers to an effectively plural and, hence, quantitative infinite. The sign of quantity, or of the fact that the *infinitum* presupposes the numerable *infinitas,* is that this *infinitas* lends itself to be thought according to the "each" of its attributes, the *unumquodque.* It is thus indubitably *composed* of undecomposable units that are its attributes. To be sure, the concept of the infinite is then under the law of difference. The infinite of attributes can be grasped in terms of the composition of the "each" but only under the pertinence of a primordial difference. The latter entails that in a certain sense one attribute is absolutely different from another. Or indeed, God's infinity is the aspect that singularizes God as Substance. It entails that He is the name of the "there is," and exposed to thought only under the sign of the multiple. The basis of this multiple is the attribute's expressive difference.

But what is an attribute? "By attribute, I mean that which the intellect perceives of Substance as constituting its essence."[5] The attribute is the essential identification of Substance through the intellect, *intellectus*. In this way the existential singularization of God is finally suspended with respect to its elucidation, or to the primary evidence of what ought to be understood by *intellectus*.

In the February 1663 letter to Simon de Vries, Spinoza painstakingly explains how the word "attribute" does not alone constitute a name of "there is" which would essentially be distinct from the name of Substance. After recalling the definition of Substance, he adds: "I mean the same thing by attribute, so long as the term is used with respect (*respectus*) to the intellect, thereby attributing to Substance the particular nature aforesaid."[6] Hence the attribute, and furthermore the multiplicity of attributes by which divine infinity is defined, is a function of the intellect. In the general apparatus of the "there is," under the name of God, there exists a singular localization, the intellect. That thought be able to open rational access to divine infinity and thus to the "there is" itself depends on the intellect's point of view or operations.

It must be agreed, then, that the intellect is in the position of a "fold," to use the central concept of Deleuze's philosophy. Were I now to use my own terminology, the intellect would be an operator of torsion. It can be localized as God's immanent production. But it is also required so that naming the 'there is' by God may be defended. For only these singular operations of the intellect lend meaning to God's existential singularization in terms of *infinite* Substance.

Just as the *clinamen* is Lucretius' enigma or the Continuum Hypothesis is Set Theory's enigma, so also is the thought of this torsion the enigma and key of the Spinozian proposal on Being.

Thinking this torsion means: How does the Spinozian determination of the "there is" return toward its inner fold, the intellect? Or more simply: How to think the being of intellect, the "there is" of intellect, if rational access to the thought of Being or of "there is" itself depends on the operations of intellect? Intellect operates, but what is the being-status of its operation?

Let us refer to Spinoza's implicit ontology—which is also the set of closure operations in his thought on Being—as all that is required to think the being of intellect. It is what the thought of a being of thought assumes as heterogeneous to the thought of Being.

The guiding thread to investigating the idea of an implicit ontology is Spinoza's construction-variation of the inner fold, or concept of *intellectus*.

The general starting point is thought (*cogitatio*) as an attribute of God. This has to do with what Spinoza calls "absolute thought," which he distinguishes precisely from intellect. "By the intellect we do not (obviously) mean absolute thought, but only a certain mode of thinking, differing from other modes, such as love, desire, etc."[7] Although it is that from which the attributive identifications of Substance exist, the intellect is itself a mode of the thought-attribute. Let us say that thought as attribute is an absolute exposition of Being, and that the intellect is the inner fold of this exposition, a fold from which there is exposition in general.

In its primary figure, the intellect is evidently infinite. It is necessarily so largely owing to its supporting role in the identification of infinitely many Substance attributes. It is the very example, and even the only example, of an immediate, infinite mode of the thought-attribute. The immediate infinite modes are described, without their existence being exemplified, in proposition 21, part I: "All things which follow from the absolute nature of any attribute of God must always exist and be infinite."[8] In July 1675, a certain Schuller asked Spinoza for examples "of things produced immediately by God."[9] Whereupon Spinoza answered that "in thought" the example is the absolutely infinite intellect.

The very concept of an infinite mode occupies a paradoxical position in the economy of Spinozian ontology. In fact, it is impossible to decide upon the existence of any such modes because neither can they be deduced a priori, nor offered up to finite experience. Let us say that an infinite mode is a well-formed concept, but existentially undecidable. Admittedly, the existence of an undecidable is only ever settled through an axiomatic assumption. This is what can be seen regarding the infinite intellect when Spinoza wrote in a letter to Oldenburg in November 1665: "I posit (*statuo*) that there exists in nature an infinite power of thinking."[10] Therefore, the infinite intellect has, if not an experimented or proven existence, then at least a status, the *status* conferred to it by a "*statuo*".

As such a statute, the infinite intellect proves to be the basis of very intertwined operations.

First and foremost, it is what gives a sense of measurement to God's power. What God may, and consequently must, produce in terms of immanent power is exactly all that the infinite intellect can conceive. "An infinite number of things must follow from the necessity of the divine nature in infinite ways—that is, everything that can fall within the sphere of (the) infinite intellect."[11]

The infinite intellect is the modal norm for the extension of modal possibility. Anything it can intellect "*omnia quae sub intellectum infinitum cadere possunt*," is expected to exist.

Clearly, we can imagine no other infinite mode having such a capacity, especially not the other examples of an immediate infinite mode given by Spinoza of motion and rest. These are apparently symmetrical to the infinite intellect from the perspective of extension. From the pure concept of motion and rest evidently no *general* prescriptions on God's power follow.

There is no mystery to the groundwork of this dissymmetry. It is due to the infinite intellect assuming a completely other *extrinsic* determination, apart from its intrinsic determination as the infinite mode of the thought-attribute. The intellect, whose components are ideas, is also determined by what it intellects or by that whose idea is the idea. This is how in addition to their affections God's attributes unreservedly comprise what is grasped, understood, or embraced (*comprehendit*) by the infinite intellect. God is that in which the intellect is located as infinite mode. This relates to the ontological relation of causality. The intellect is an immanent effect of God. But the intellect is also such that God and his attributes are what it understands, or at least the correlate of ideas constituting It. For every idea is an 'idea of . . . ' and is correlated to an *ideatum*. Or, there is an object of the idea. In this sense, God's attributes and the modes of these attributes are the objects of the infinite intellect.

The notion of object for an idea is all the stronger for Spinoza to expressly declare that the object partly singularizes or identifies the idea, especially as far as what he calls "reality" is concerned. Hence, the note to proposition 13, part II: "Still, we cannot deny that ideas, like objects, differ one from the other, one being more excellent than another and containing more reality, just as the object of one idea is more excellent than the object of another idea and contains more reality."[12]

Causality put aside, this clearly assumes *a second fundamental relation*. It is a relation that only makes sense for the intellect and singularizes it absolutely. We know that for Spinoza, who is in no way an empiricist, the relation between the idea and its ideatum, that is, the idea and the object of the idea, never arises from causal action. This is the meaning of proposition 2, part III: "The body cannot determine the mind to think, nor the mind the body to motion nor to rest."[13] There is no conceivable causal relation between the idea and its object, for the relation of causality works strictly within an attributive identification. Whereas, and this is where the whole problem lies, the object of an idea of the intellect can very well be a mode for an attribute other than thought.

So as to span the disjunction between attributes, a special relation is needed. This relation cannot be causality. I name it "coupling" *(couplage)*. An idea of the intellect is always coupled to an object, which means that a mode of thought is always coupled to another mode that might be one of extension, thought, or any other attribute.

That this is a strong relation is attested to by the fact that Spinoza at no time hesitates to call it a "union." "That the mind is united with the body we have shown from the fact, that the body is the object of the mind; and so for the same reason the idea of the mind must be united with its object, that is, with the mind in the same manner as the mind is itself united to the body."[14] This is how it might be seen in general that there is union between the idea and its object, which includes a span over the disjunction between attributes. It is this union, the radical singularity of the intellect's operations, that I call "coupling."

Plainly, we have to add that coupling has a norm. An idea is more or less "well coupled" with its object. A completed coupling is called a "truth." This is what is said as of axiom 6, Part I: "A true idea must agree with that of which it is the idea or ideatum."[15] Agreement is what norms the coupling and makes a truth of it. This norm of agreement, just like the coupling relation, is extrinsic. It is not, as with causality, strictly immanent to the attributive determination. In his unpacking of definition 4, part II, Spinoza takes care to distinguish an intrinsic norm of the true, that is, likeness *(adéquation)* that ultimately refers to causality, from an *extrinsic* denomination of the true idea, which is "the agreement between the

idea and its ideatum (that of which it is the idea)."[16] Agreement refers
not to causality, but to coupling. It is clear that no other infinite mode
than the intellect demands of the terms constituting it to bear a coupling relation, as the case is with the idea. Still less do the other
assumed infinite modes have to adapt to the coupling norm, that is,
the agreement (*convenance*) whose result is called "truth."

The relation of coupling, like the causal relation, prompts the
existence of infinite series. Every mode has a cause, which in turn
has a cause, so on and so forth. Likewise, every idea coupled to its
object is compulsorily in turn the object of an idea coupling with
it. This is the well-known theme of the idea of the idea, examined
in closer detail in the note to proposition 21, part II, in the peculiar
case of the mind as idea of the body, and the idea of the mind as
idea of the idea. The text subtlety blends ontological identity and
the coupling relation: "The mind and body are conceived as one and
the same individual, conceived now under the attribute of thought,
now under the attribute of extension; wherefore the idea of the mind
and the mind itself are one and the same thing, which is conceived
under one and the same attribute, namely, a thought. [...] Strictly
speaking, the idea of the mind, that is, the idea of an idea, is nothing
but the distinctive form of the idea in so far as it is considered a mode
of thinking without reference to the object."[17] The "one and the same
thing" seems to make every underlying difference to the coupling
relation disappear. Yet this is not exactly the case. For it is only the
couple, as the intellect captures it, that identifies the individual. The
upshot is that the idea of the body, in as much as it couples by
straddling the attributive disjunction, remains necessarily distinct from
the idea of this idea, which is immanently coupled to the thought-
attribute. We can also put it like this: beneath the relation, there is
always an identity effect. This is the same individual diversely
intellected either as body or as mind. It is the very mind twice
intellected. But this identity effect is intelligible only in the categories
of the intellect, which precisely emerge from the coupling.

Finally, the active structure of the infinite intellect is radically
singular. It proves to be exorbitant to the general principles of ontological naming, which are as follows:

• It depends on the undecidability tied to the infinite modes.

• It measures up the total power of God.

- It imposes another relation in addition to causality, namely, coupling, which subverts identity fields.

- It bears at each of its points or ideas not only one infinite recurrence according to causality, but two of them according to coupling.

In fact, the infinite intellect is itself an exception to the infamous proposition 7 in part II: "The order and connection of ideas is the same as the order and connection of things."[18] For in no attribute other than thought is a structure conceivable—that is, representable by the intellect—which is isomorphic to the structure of the intellect itself. Therefore the thought-attribute is not isomorphic to other attributes, even according to the relation of causality.

Turning now to human intellect, or finite intellect, things get even more complicated.

The main difficulty is as follows: Can the finite intellect be conceived as a modification or affection of the infinite intellect? After all, this is what seems to impose the causal relation as constitutive in the immanent determination of the "there is." Unfortunately, the answer to the question is no. For proposition 22 part I ascertains that "all that follows from an attribute of God, in so far as it has been modified from one such modification, and that by this attribute it exists necessarily and as infinite, must also exist necessarily and be infinite."[19] Let us make this clearer: everything that follows from an immediate infinite mode, like the infinite intellect, is infinite. There is therefore no way that the finite intellect could be an effect of the infinite intellect. Why do they bear the same name, then?

To solve this problem, Spinoza proposes, not without a tinge of timidity, a *third* fundamental relation subsequent to causality and coupling, which may be called "inclusion." To be sure, the finite intellect is not an effect of the infinite intellect. But, as Spinoza tells us, it is a *part* of it. This is what is stated, without really being a proof or elucidation of the concept at stake, in the corollary to proposition 11 in part 2: "The human mind is a part of God's infinite intellect."[20] This completely novel inclusive relation actually deals with what proves to be the major difficulty in Spinoza's ontology: the relation between the infinite and the finite.

That it is a question of inclusion, namely, a set-related vision, is borne out by its reciprocal statement. Just as the finite intellect is a part of the infinite intellect, so also is the infinite intellect the assembly or collection of finite intellects. "Our mind, in so far as it understands (*intellecte, intelligit*), is an eternal mode of thinking determined by another eternal mode of thinking and this other by a third, and so on to infinity; so that all of them taken together at once constitute the eternal and infinite intellect of God."[21] By summing an infinite chain of finite modes to the infinite, the infinite intellect proves to be the *limit point* of the finitudes it totalizes. Conversely, the finite intellect is a composition point for infinite summing. Causality is but apparent order here, as it remains powerless in the task of leaving the finite. For, as proposition 28 in part 1 ascertains, a finite mode never has another finite mode as its cause. The true relation here is inclusive.

Spinoza does not shy away from saying elsewhere everything he reckons to be wrong with an uncontrolled handling of the relation between the parts and the whole. As for the intellect, and justifying the fact that the same word can designate both human operations and the thought-attribute operations of the inner fold, he has no choice but to go in that direction. Inclusion alone globally explains the *being* of the finite intellect.

Were we now to seek what the *operations* of this intellect might be, we would immediately come upon the coupling relation. Its essential motif is to identify the human mind through its coupling with the body. Thereupon we eschew direct involvement with the third, inclusive relation by remaining at the local level, as it were. The human mind is an idea, hence a finite component of what the infinite intellect names "superior modality." It is the idea of the body.

The great advantage of this purely local treatment is to explain all things obscure in terms of finite thinking. Let us be reminded about how coupling has agreement as its norm. If the idea does not agree with the object with which it is coupled, it is obscure and not true. The whole obscure side of thought is generated and measured by the norm of fitting. The key is found in proposition 24, part II: "The human mind does not involve adequate knowledge of the parts composing the human body."[22] This is said even more brutally in the proof to proposition 19, part II: "The human mind has no knowledge of the human body."[23]

Let us take note of the complexity here. Ontologically, the mind is an idea, the idea of the body. But this does not imply that the mind understands its object. Coupling between the idea and the object of the idea is subject to degrees of knowledge. It is more or less subject to the agreement norm. This is the case even more so in that it has to do with a complex idea referring to the body's multiple composition.

Finally, the ontology of finite intellect, at the cost of having to use a third relation, inclusion, explains all of the themes in Book V. As we leave off from the infinite intellect, we start experimenting with being eternal. The theory of the intellect's operations, at the cost of using the second, coupling relation makes the themes in Books III and IV clear: we do not immediately have an adequate idea of what our own intellect might be.

The link between the two is certainly not simple. In point of fact, the difficulty becomes the following: if the finite intellect is defined as an ideal coupling to the body without adequate knowledge of its object, whence can it have true ideas? To be sure, the inclusive relation explains it, but it is only a global metaphor. What is the *local* operation of truths?

The problem is not knowing whether we have true ideas in the extrinsic sense of the agreement norm. For we certainly experience such ideas. The true idea bears itself, including its validation by coupling and agreement. The note to proposition 43, part II, unpacks this well-known theme: "How can a man be sure he has ideas that agree with their objects, and that his knowledge arises from the simple fact of having an idea that agrees with its object—in other words, that truth is its own norm?"[24] Here Spinoza insists on unifying the operative approach in coupling and the purely ontological approach in including. He goes on to write: "We may add that our mind, in so far as it perceives things truly, is a part of the infinite intellect of God."[25] Accordingly, existence of true ideas is globally ensured by inclusion of the finite intellect by the infinite intellect locally through the express exposition of the agreement of a coupling.

The real problem is: How? How does the finite intellect have true ideas, given that it does not have adequate knowledge of the body-object whose idea it is?

The solution to this strictly operational problem, since it is not existential, is elaborated in propositions 38 and 40, part II. The latter ascertain that every idea referring to a property common to all bodies, or all ideas, indeed to all there is in so far as it is, is necessarily true. And that the ideas following from true ideas are equally true.

In other words, there is no true knowledge of the singular body of which our mind is the idea. But regarding what is common to all bodies and consequently not singular, there is a true idea necessarily in the finite intellect once the latter can be coupled with it.

We have true ideas because the finite intellect possesses ideas coupled to nonsingular objects, to *common* objects.

Finally, veridical reason is woven from the common notions.

Spinoza's constant polemicizing against universals and homonyms devoid of a being-content is well known. In a way, his doctrine only accepts existence of singularities in terms of the immanent effect of the divine "there is." From another angle, the only acceptable proof for a local operation of true ideas stands entirely with the common notions or generic properties of singularities. The true is generic, whereas Being is the power of singularities.

Spinoza has no qualms about repeating this. "The notions called common are the fundaments of our deductive capabilities," he states.[26] And more firmly still in the proof to corollary 2, proposition 44, part II: "The fundaments of reason (*fundamenta rationis*) are the notions, which explain things common in all, but do not explain the essence of any singular thing; for this reason, they must be conceived without any relation to time, though under a certain kind of eternity."[27]

The objection according to which the third genus of knowledge would essentially be distinct from reason, thereby opening 'lateral' (or purely intuitive) access onto singularities themselves, does not hold. It is too old and complex a debate to go into detail here. Let us observe merely how the preface to Part V identifies in an altogether general fashion the "power of mind" with "reason": "*de sola mentis, seu rationis potentia agam.*" And that if the third genus of knowledge is indeed an "intuitive science," as in the famous formula of the "eyes of the soul are the proofs themselves," then an "intuition" of that eye is obviously tantamount to grasping the proofs "in one shot," or in an instantaneous glance of the deductive bond between the common notions. Still, this does not manage to release us from the pure universality in which the true ideas of finite intellect are quartered.

This is how we are led back to the pure eternity-based axiomatic from which we left off. If the field of the thinkable—for a finite intellect—is gauged by "what is common to all," it is really a case for the apparatus of the "there is," namely, the attributive identification of the divine infinite.

This circular closure of Spinoza's ontology through a mediation of the structures of the intellect is hatched from a series of complex schemes. Let us recapitulate them.

1. It is only possible to open the path to identifying the "there is" under God's name by previously understanding difference. In turn, this upholds the purely extensive conception of the divine infinite.

2. The possibility of the extensive conception of the divine infinite presupposes an internal fold for both the attributes and gauge of divine power, namely, an irreducible singularity: the infinite intellect.

3. The infinite intellect has all of the characteristic features if not of a Subject, then at least of subjective modality or the predicative power linked to this effect. As an immediate infinite mode, it is inaccessible to the usual paths by which existence is ascertained. It is thus existentially undecidable. In its structure, the infinite intellect requires a foreign relation to the sole relation initially proposed, which is causality. This second relation is coupling. It has a norm of agreement, by which the truth is gauged. Let us say that the intellect's operation, as a truth operation, is atypical. Finally, coupling renders every point of the intellect infinite, just like causality renders every point of the "there is" infinite. Let us say that the intellect is intrinsically a double of immanent productive power.

    The way I see it, these features—the undecidability of existence, the atypical characteristic of the operation, and the doubling effect—identify the intellect as a modality of the subject-effect.

4. The human intellect, or finite intellect or mind, in turn demands a third relation, inclusion, so as to be localized. Just as the coupling relation makes way for overstepping the disjunction between different attributes, so also does

the inclusion relation make way for overstepping the disjunction between the finite and infinite. The intellect is then ontologically determined as a local point of the infinite intellect, which is the recollection of all these points. If we agree that the infinite intellect is the intrinsic modality of a subject effect, then the human intellect can be said to be a localized subject effect. Or it is a differential subject. Or quite simply: a Subject.

5. The human intellect can also be defined by coupling. The immediate upshot of this is that the only points of truth are axiomatic and general. The singular is subtracted from every local subjective differential. This can also be said as: the Subject and the human mind's sole capability for the true is a mathematics of Being, or Being as mathematically thought. Every truth is generic.

Or else: what can be thought of Being is mathematical.

After which, I conclude that the *more geometrico* is true thought itself, as the thought of Being or of the "there is." Being can be thought only *more geometrico*. Conversely, all mathematical thought is thought of Being in a finite localization. As a matter of fact, this is why "the eyes of the mind are the proofs themselves." Outside of mathematics, we are blind.

In my opinion, the outcome is plain to see. God must be understood as mathematicity itself. The name of the "there is" is "*matheme*."

However, and in Spinoza's text itself, the paths by which to ascertain this result constrain us to open a space of thinking which is not under the norm of naming the "there is." These are what I call "operations of closure." Constitutive terms of this space are indeterminacy, difference, Subject, undecidability, atypicality, coupling, doubling, inclusion, and the genericity of the true, as well as a few others.

In the task of making this underbelly of mathematics explicit, we lack a founding category that is either exempted from the "there is," or supplements it. This is the point at which the concept of what I call the "event" has to enter. By introducing it, I am only treading on many others' footsteps. The event is also what grounds time, or rather times, event by event. Spinoza cared little for this. He strove to think according to his own expression "with no relation to time." He saw freedom expressed in "constant and eternal love of God."

We can say instead, love in the pure elevation of the matheme, or better yet, in the love of the "there is," a kind of "intellectual" love that is never but the intuitive abbreviation of a proof, a glance with the mind's eyes.

Other thoughts do open from within the very doubling of this exclusive thought. These thoughts take on the mathematics of being-multiple. They are explicitly Spinozian from this angle. But their real thrust draws instead from implicit and paradoxical Spinozism, that is, from the event torsion according to which the paradox of a Subject under the name of "intellect" befalls.

These thoughts practice an elevation of the matheme. But wary of what exceeds or grows impatient with it, they no longer consent to grant divine names to the matheme.

This is why they have access to infinity without being bogged down by finitude. So they meet up with a more Platonist than Spinozian inspiration.

Platonism is the big question. I have been holding it back by indicating that Plato both discovered and underestimated the scope of mathematics as a science of Being *qua* Being. I also said that only at a later point would I take the straight and narrow Platonist path against the empirical (Aristotelian) vision of mathematics. That time has now come.

6

PLATONISM AND MATHEMATICAL ONTOLOGY

In the introduction to *Philosophy of Mathematics,* an anthology edited by P. Benaceraff and H. Putnam, we read the following:

> Platonists will be those who consider mathematics to be the discovery of truths about structures that exist independently of the activity or thought of mathematicians.[1]

This criterion of the exteriority (or transcendence) of mathematical structures (or objects) identifies how "Platonism" is conceived in virtually all works in the "philosophy of sciences." Now, this identification is surely inexact. Its inexactness is the outcome of a distinction that the "Platonist" presupposes between interiority and exteriority, between the knowing subject and the known "object." In fact, this distinction is absolutely foreign to the real Platonic apparatus. As anchored as this distinction might be in current epistemology, as ascertained as the theme of object and objectivity in view of a Subject and the subjective actually is, it stands to reason that from the basis of such presuppositions one can only end up entirely missing the process of thought at work with Plato.

The "independent existence" of mathematical structures should be highlighted as something completely relative for Plato. What the metaphor of reminiscence refers to is precisely that thought is not, or indeed is never, confronted to "objectivities" from which it would be separate. The Idea is always already posed, here and now. If it were not "activated" in thought, it would remain unthinkable. As for mathematical ideas more specifically, the whole point of the concrete demonstration in the *Meno* is to ascertain their presence in the thinking of the least educated and most anonymous mind, none other than the slave's.

Plato's fundamental concern is to declare the immanent identity and co-belonging of the known and the knowing mind, and their essential ontological commensurability. Were there a point on which he is Parmenides' son, it would be when asserting, "the same is thinking and Being."[2] In so far as mathematics touches upon Being, it is intrinsically a thought. Reciprocally, if mathematics is a thought, it touches upon Being itself. The motif of a knowing Subject that would have to "aim" for an outer object—a motif whose background is empiricist even when the object is assumed to be ideal—is entirely inappropriate to the philosophical usage Plato makes of the existence of mathematics.

Plato is concerned so little with mathematical structures existing "in and of themselves" that:

1. Ideality is in fact the general name of what befalls to the thinkable and in no way singularizes mathematics. As old Parmenides once commented to young Socrates, in as much as we think mud or hair, one must grant that there is the idea of mud and the idea of hair. In fact, "Idea" is the name of what is thought *as* thought. The Platonic theme consists precisely in making immanence and transcendence indiscernible. It seeks to settle in a thinking space where this distinction is inoperative. A mathematical idea is neither subjective ("the mathematician's activity") nor objective ("structures existing independently"). It is in one flail swoop a break with the sensible and a position of the intelligible, something that one ought to call a thought;

2. Plato is not interested in the status of the purported mathematical "objects," but by the movement of thought. That is because mathematics is summoned in the end only to identify dialectics by means of difference. Admittedly, everything in the thinkable is Idea. It is thus in vain to seek on objectivity's side some such difference between regimens of thought. The singularity of movement alone (leaving off from hypotheses or treading back to first principles) authorizes one to delimit mathematical *dianoia* from

dialectical (or philosophical) intellection. The separation of "objects" is second and always obscure. It is an out-sourcing "in Being" of indices drawn from thought.

Finally, one thing alone is sure: Mathematics is a thought (which, in Plato's parlance, means that it breaks with the sensible immediate). Dialectics is also a thought. And when considered in the protocol of their actual practice, these two thoughts are different thoughts.

From this idea, we can attempt to draft a definition of the Platonic inscription of the mathematical condition of "philosophizing":

Platonism is the recognition of mathematics as a thought that is intransitive to sensible and linguistic experience, and dependent on a decision that makes space for the undecidable, while assuming that everything consistent exists.

To size up the polemical extension of this definition of Platonism, let us turn toward the one proposed by Fraenkel and Bar-Hillel in *Foundations of Set Theory:*

A Platonist is someone convinced that, in association with every well-defined monadic condition [i.e. in which a predicate is attributed to a variable, like P(x)], there usually exists a set or class that includes all of the entities satisfying this condition, and them alone; and in turn, this set is an entity whose ontological status is similar to that of its elements.[3]

I myself do not believe a Platonist can be convinced in such a way at all. Plato himself constantly strives to show that the correlate of concepts or well-defined propositions can be empty or inconsistent. Or that the "entity" corresponding to it can require an ontological status that is exorbitant to everything involved in the starting formula. This is how the correlate of the Good, as clearly definable as the notion might be and as evident as its practical instance is, requires a dispensation in Being from the status of the Idea—the Good lies "beyond" the Idea. The explicit purpose of the *Parmenides,* with such perfectly clear statements as "One is" and "One is not," is to ascertain that, beneath any assumption whatso-ever regarding the correlate of the One or of "others than (the)

One," an inconsistency is obtained. After all, this is the first example, though a purely philosophical one, of an absolutely undecidable argument.

Against A. A. Fraenkel and Yeroshua Bar-Hillel's declaration, I maintain that the undecidable is a crucial category for Platonism. The point is precisely that one can never predict whether there corresponds a thinkable entity to a well-defined formula. The undecidable attests to a Platonic lack of trust in language's clarity to be able to decide upon whether something exists. In this sense, it is Zermelo's axiom that ends up being Platonist. For a given formula, it grants the existence of 'entities' that validate it as it also grants the collection of such entities provided they be in a formerly given existing set. For thought requires a constant and immanent guarantee of Being.

As for the content of Plato's dialogues, the undecidable commands the perplexing aporetic style of the dialogues. This course leads to the point of the undecidable so as to show that thought precisely ought to decide upon an event of Being; that thought is not foremost a description or construction, but a break (with opinion and/or with experience) and, therefore, a decision.

In that regard, Gödel appears to me to have been more lucid. After all, he is always ranked by "the philosophy of mathematics" as among the "Platonists."

Consider the following passage taken from his celebrated essay, "What is Cantor's Continuum Problem?":

> However, the question of the objective existence of the objects of mathematical intuition (which, incidentally, is an exact replica of the question of the objective existence of the outer world) is not decisive for the problem under discussion here. The mere psychological fact of the existence of an intuition which is sufficiently clear to produce the axioms of Set Theory and an open series of extensions of them suffices to give meaning to the question of the truth or falsity of propositions like Cantor's continuum hypothesis. What, however, perhaps more than anything else, justifies the acceptance of this criterion of truth in Set Theory is the fact that continued appeals to mathematical intuition are necessary not only for obtaining unambiguous answers

to questions of Transfinite Set Theory, but also for the solution of the problems of finitary number theory (of the
type of Goldbach's conjecture), where the meaningfulness
and unambiguity of the concepts entering into them can
hardly be doubted. This follows from the fact that for every
axiomatic system there are infinitely many undecidable
propositions of this type.[4]

What are the most important features of this "Platonic" essay?

- The word "intuition" has no other sense than that of a decision on inventive thinking with respect to the intelligibility of
axioms. According to Gödel's own expression, it is a case of
being able to "produce the axioms of Set Theory," and the
existence of this capacity is a "simple fact." Let us recall that
the intuitive function does not go the mile in grasping "outer"
entities, but clearly decides upon a first or irreducible proposition. The comprehensive invention of axioms is what vouches
for the mathematical proposition as thought and, consequently,
exposes it to the truth.

- The question of the "objective" existence of purported objects is explicitly declared to be secondary (it "is not decisive
for the problem under discussion here"). In addition, it does
not characterize mathematics because this existence is ac-
commodated beneath the same banner as is the outer world.
In effect, it is very Platonic to see nothing more nor less in
mathematical existence than in existence proper. In all cases,
the thinkable (mud, hair, a triangle or complex numbers) can
be questioned as to its existence, which is something other
than its being. As for its being, it is attested to solely in virtue
of being embodied by a thought.

- The crucial problem is truth. Once there is intuitive thinking
(and the intelligibility of the axioms records that fact), one can
"give meaning to the question of the truth or falsity" of propositions authorized by this thinking. This sense stems precisely
from the fact that the thinkable, in terms of the Idea, necessarily touches upon Being. "Truth" is never but the name by
which Being and thought match up in a single process.

- The infinite and finite do not constitute a paramount distinction for thought. Gödel insists on how "acceptance of this criterion of truth" is the outcome of intuition (or the axiomatic decision), which is continually required either for breaking with the problems of finitary arithmetic as well as problems concerning transfinite sets. This is to say the movement of thought, which alone concerns us here, is not essentially different in the infinite than in the finite.

- The undecidable is organically bound to mathematics. It is not so much the case of a "limit," as sometimes claimed, than a perpetual incentive to practice inventive intuition. From the idea that every mathematical thought apparatus recapitulated in the founding axioms contains an undecidable, we may conclude that intuition is never useless. Mathematics has to be constantly decided over and over again.

Finally, I should like to characterize in three points what it is legitimate to call a Platonic philosophical orientation with respect to the modern mathematical condition and, consequently, with respect to ontology.

1. Mathematics is a thought.

I have already developed this assertion at length. It is so important that I would like to at least punctuate it again. As an example, let us recall L. Wittgenstein, who by no means lacked knowledge on the matter, when declaring that "the mathematical proposition expresses no thought."[5] All he does is take up again in customary radical fashion a major overall thesis of empiricism and sophistry. We shall never finish up with refuting it.

That mathematics is a thought means in particular that, with respect to it, the distinction of a knowing Subject and a known object has no pertinence. There is a regulated movement of thought, coextensive to the being it embodies—which Plato named "Idea." It is a movement wherein discovery and invention are strictly indiscernible, in the way the idea and its ideatum are indiscernible.

2. Every thought—and therefore, mathematics—sets off decisions (intuitions) from the standpoint of the undecidable (of nondeducible inference)

The outcome of this feature is a maximum extension of the principle of choice regarding the thinkable. Since decision comes first and is continually required, it is vain to attempt to reduce decision to constructive protocols or to those whose norms come from without. The constraints of construction (often misnamed "intuitionistic" constraints, for the true advocate of intuition is the Platonist) ought to be, on the contrary, subordinated to the liberties of thoughtful decision. This is why the Platonist sees nothing to change, provided the effects of thought be maximal, regarding free usage of the law of the excluded middle and consequently of the *reductio ad absurdum*.

3. Mathematical questions on existence refer only to the intelligible consistency of what is thought

Here, existence has to be considered an intrinsic determination of effective thought insofar as it embodies Being. That existence does not embody Being is always borne out through an inconsistency, which must be carefully distinguished from undecidability. In mathematics, Being, thought and consistency are one and the same.

From these features important consequences arise. Modern Platonism can be deciphered in them, a Platonism of manifold-being.

- Foremost, as Gödel indicated, the indifference to the purported "paradoxes" of the actual infinite. In as much as the sphere of intelligibility instituted by the infinite manifestly poses no specific problems, neither in the axiomatic intuition nor in the proof protocols, the grounds for worry are always extrinsic, psychological or empiricist. Mathematics is denied its self-sufficiency regarding the regimen of the thinkable it determines.

- Next, the desire for maximality in granting existence, namely, the greater amount of existence, the better it is. The Platonist handles the audacity of thought. She is repugnant to restrictions and censures issued from without (and in peculiar timorous philosophemes). As much as thought owes to the being it embodies not to sink into inconsistency, one can and ought to go ahead with assumptions of existence. This is how thought follows a line of intensification.

• Finally, recognition of a criterion when an apparent option is imposed onto the becoming of mathematics. This criterion is precisely that of maximal extension for the thinkable consistent entity. Thus, the Platonist will concede the Axiom of Choice instead of its negation, for the universe with Axiom of Choice is otherwise broader and denser in significant links than is one without the axiom. Conversely, the Platonist will be more reserved about any concessions made on the Continuum Hypothesis and, even more, on the hypothesis of constructibility. For the universes regulated by these hypotheses appear narrow and constrained. The constructible universe is even peculiarly cunning. F. Rowbottom has shown that if one allows a peculiar kind of large cardinal (the Ramsey cardinal), constructible real numbers are then denumerable. A denumerable continuum appears to the Platonist as too-convincing an intuition. Rowbottom's theorem comforts the Platonist in his conviction: give precedence to decided consistencies over controlled constructions.

I have shown Cantor's conceptions clarify the thought of Being as pure multiplicity. A set-theoretic decision regarding mathematics involves its ontological repetition. It may now be observed to impose a Platonic orientation, in the sense I have just laid out. Besides, this is what confirms Gödel's philosophical choices, which established him as the greatest of Cantor's continuators (along with Cohen).

The fact of the matter is that Set Theory is the perfect example of a theory wherein (axiomatic) decisions win out over (definitional) constructions. Moreover, empiricists and advocates of the "linguistic turn" of our times have not failed to object to the theory. They did not even manage to define or elucidate its organic concept, the set. To which a Platonist like Gödel would retort that what counts are the axiomatic intuitions constituting a truth space, and not the logical definition of primitive relations.

Unlike the Aristotelian orientation (potentiality as the first singularization of substantiality) or the Leibnizian one (logical possibility as "pretension to Being"), Set Theory knows only the actual manifold. That actuality be the effective form of Being and that possibility or potentiality be fictions—both of these are profoundly Platonic motifs. Nothing is more significant in this regard than the set-theoretic treatment of the concept of function. What appears to be a dynamic operator, often borne by spatial or physical schemes, (if $y=\smallint(x)$, y will be said to "vary" as a function of the variations of x, etc.), is strictly dealt with in the set-theoretic framework as an actual manifold. The function's manifold-being is its graph, that is, a set whose elements are ordered pairs of the (x,y) type. Any allusion to dynamics or "in variation" is eliminated.

Likewise, the concept of "limit," so affected by the experience of becoming, tending-toward and asymptotic movement, is brought back to the immanent characterization of a certain type of multiplicity. Thus, to be identified, an ordinal limit does not have to be represented as that toward which the chain of ordinals "tends," or of which it is the limit. The reason is that it *is* this series itself (that the elements of this series are what define it as a set). The transfinite ordinal χ_0 that comes "after" the natural numbers is nothing but the set of all natural numbers.

In clear affiliation with the Platonic spirit, virtuality is everywhere conceived as actuality. There is but one type of Being, the Idea (or, here, the set). Therefore, there do not exist actualizations. For any kind of actualization assumes that several regimens of existence be granted (or, at least two, potency and the act).

Besides, Set Theory abides by the principle of existential maximality. As early as Cantor, its inspiration was to go beyond all foregoing limitations, all criteria—held as extrinsic—of 'reasonable' existence. Granting admission to increasingly gigantic cardinals (inaccessible, Mahlo, measurable, compact, supercompact, enormous cardinals, etc.) fits its natural genius. On the other hand, surreal number theory admits infinitesimals of all sorts. In addition, this apparatus deploys increasingly saturated and complex "levels" of Being, an ontological hierarchy (cumulative hierarchy) that, in line with an intuition now showing a strong Neoplatonist accent, is such that its (inconsistent) "totality" is always reflected consistently at one of its levels in the following sense. If a statement is valid "for the whole universe" (in other words, if the quantifiers show no limit, if "for all x" means "for some such set of the whole universe"), then there exists a set in which this statement is valid (the quantifiers this time taken as "relativized" to the set at stake.) What this means is that this set, considered as a "special universe," reflects the universal value of the statement and localizes it.

This reflection theorem also tells us that what can be uttered with respect to "limitless" being can also always be uttered in a place. Or even that every statement prescribes the possibility of a localization. The Platonist theme of the intelligible localization of the rational Word per se can be recognized here. This is identical to what Heidegger criticized as the operation of the Idea "cutting up" the "coming into unconcealedness" of the natural beingness of Being.

More to the point, Set Theory's Platonic vocation bears on three categories constitutive of any philosophical ontology: difference, the primitive name of Being and the undecidable.

For Plato, difference is settled by the idea of the Other. Admittedly, the way this idea is presented in the *Sophist* necessarily implies an intelligible localization of difference. Insofar as the idea "participates" in the Other, it can be declared different from another idea. There is a localizable evaluation of difference: the very way in which an idea, albeit "identical to itself," participates in the Other as another idea. In Set Theory, this point is broached by the Axiom of Extensionality: if one set is different from another, the case is that there exists at least one element belonging to the one and not the other. This "at least one" localizes difference and prohibits purely global differences. There is always a point of difference (just as with Plato there is an idea that is not "other" than another "in and of itself," but solely in as much as it participates in the Other. This is a capital feature, especially because it limits the (Aristotelian, as well as Deleuzian) rights of qualitative, global, and natural difference.

In Set Theory, the primitive name of Being is the void, the empty set. The whole hierarchy takes root in it. In a certain sense, it alone "is." And the logic of difference implies that the void is unique. Indeed, it cannot differ from another since it contains no element (no local point) that can aver this difference. This combination of primitive naming by the simple absolute (or the in-different, which is the status of the One in the *Parmenides*) and of a foundational unity is indubitably Platonic. The existence of what covers this primitive name (namely, the existence of the empty set) has to be axiomatically decided, just like—this is the sense of the aporia in the *Parmenides*—it is vain to want to deduce existence (or inexistence) from the One: we have to decide and accept the consequences.

Finally, as we have known since Paul Cohen's theorem, the Continuum Hypothesis is intrinsically undecidable. Many believe Cohen's discovery has driven the set-theoretic project into ruin. Or, at least it has "pluralized" what was once presented as a unified construct. I have discussed this enough elsewhere for my point of view on this matter to be understood as quite the opposite.[6] What the undecidability of the Continuum hypothesis does is complete Set Theory as a Platonist orientation. It indicates its line of flight, the aporia of immanent wandering in which thought experiences itself as an unfounded confrontation with the undecidable. Or, to use Gödel's lexicon: as a continuous recourse to intuition, that is, to decision.

The anti-qualitative localization of difference, unity of existence under a primitive naming, intrinsic trial of the undecidable: such are the features according to which Set Theory can be grasped by philosophy from the perspective of a truth theory that lies beyond a mere logic of forms.

The objection can be raised that, in its proof protocols, all of mathematics is indeed sustained by a logical substructure. How, then, is mathematics articulated as a thought of Being *qua* Being or pure manifold theory, and how is mathematics articulated as a "formal" science in terms of remaining bound to proof protocols?

To enter into this redoubtable question, which deals with the ontological status of logic and the structure of the onto-logical, it is useful to return to the Aristotelian conception. After all, the latter is Platonism's primitive "other."

7

THE ARISTOTELIAN ORIENTATION AND LOGIC

The heart of any "Aristotelian" relation to mathematics is to consider mathematics as not a thought. We have seen that Aristotle himself, who most likely was not Aristotelian, nonetheless concluded that mathematics ultimately is not a matter of ontology, but of aesthetic satisfaction.

In this sense, our century is much more Aristotelian than we imagine. Besides, this is an ineluctable effect of the century's essential anti-Platonism, whose prophet is Nietzsche. Yet that stance is shared as much by the Anglo-American "linguistic" orientation, which incessantly vituperates against "naive Platonism," as it is also by the Heideggerian hermeneutic orientation, in the view of which Platonism deleted the original "unconcealing" of Being as *physis* under the Idea. Even the former-USSR's science dictionary insisted on emphasizing the materialist merits of Aristotle, whereas it treated Plato as the slave owners' ideologue. All of this merely stresses the scope of the consensus.

What does the assertion according to which mathematics is not a thought actually mean? What it obviously does not mean at all is that it fails to constitute a coherent and rational knowledge domain. Instead, it points to the idea that if left unanchored to a principle of Being this knowledge domain cannot profess truth. Under these circumstances, it matters little whether the "principle of Being" invoked is of a metaphysical or empirical nature, like Aristotle's substance and Leibniz's monad or sense data for the positivists, respectively. In either case, the central thesis is that mathematics remains purely formal (or "void of Being"), which bars it from the real embodiment required by any actual thought.

For a Platonist, the Idea or Form—whatever the ontological status finally assigned to this term might be—explicitly designates the knotting of mathematics to a real. It is what makes sense of the idea that we can talk of mathematical truths. For an Aristotelian or Leibnizian, categorizing Being beneath the species-type of a singularity (that is, substance as matter's local information, or the monad as a "metaphysical point") unfastens mathematics from any real inscription. That is because the triangle or differential is neither a substance nor a monad.

Whether mathematics is or is not a thought has no bearing as a judgment on its importance *for* thought. It is well known that Leibniz's metaphysics is entirely sustained by his mathematical conception of the continuum, that is, calculating maxima, and so forth. Mathematics is certainly more important for the edification of Leibniz's system than for Plato's aporetic ontology. To a certain extent, Aristotle's dealings with mathematics are more precise than Plato's, but many things that are not thoughts are of paramount importance to thought. For both Leibniz and Aristotle, mathematics is woven only of purely ideal, if not fictitious, relations. It supplies the conventions of the virtual intelligible. It has to do with the art of computation. It is an art grounded in reason, though in no way does it make a thinkable dent in Being.

More precisely, mathematics is a grammar of possible existence. This point is surely a decisive one. For the Platonist, mathematics is a science of the Real (which is moreover Lacan's own entirely Platonic definition.) For Aristotle or Leibniz, mathematics recapitulates certain formal givens of possible-being. These givens are essentially analytic, meaning they do not touch upon the forever synthetic singularity of what is.

For a Platonist, thought is never descriptive. It is what arises from a break with description, since it is intransitive to opinion and hence to experience. For an Aristotelian, thought is the construction of an adequate descriptive framework in which experience and opinion discover ways—without succumbing to any breaks—of being grounded in reason. Nothing is more striking than the stylistic difference involved in these representational differences of thought. For a Platonist, what counts are rupture principles. For an Aristotelian, what counts are legitimation protocols. When this opposition is applied to mapping mathematics onto philosophy's field we get the following: a Platonist's entire interest focuses on axioms in which the decision of thought is played out, whereas an Aristotelian's (or Leibnizian's) interest focuses on definitions in which the representation of possibilities is laid out.

What this is all amounts to is really quite simple. For either an Aristotelian or Leibnizian, the essence of mathematics is logic. There is no coincidence that Aristotle was the one to have authored the *Second Analytics,* which is history's first formal logic. Nor is there anything striking about the fact that, ever since his youngest days, Leibniz worked on a "universal characteristic," which grants him the merit of being considered the ancestor of modern mathematical logic. For these thinkers, mathematics works in terms of coherent possibilities. Deprived of an ontological basis, it abstractly idealizes the admissible consecutions, that is, the control algorithms of a "true" thought appropriating singularities, whether that thought be substantial or monadic. Mathematics is therefore the general logic of what is rationally possible.

If mathematics is a logic of the possible, then questions of existence are not intrinsic to it (as they are for a Platonist). The fundamental problem that philosophy poses regarding mathematics then no longer has to do with its movement of thought and its linkage to Being. Let us say it is no longer about questioning its truth since the purely ideal dimension of mathematical entities is accepted here. Rather, the problem lies with the empirical, linguistic and rational *provenance* of mathematical idealities. So a slope is needed by which to ascertain this provenance and thereby prevent their forms from having too much license, or to be unduly taken as truths.

What allows us to ascertain the provenance of ideal mathematical notations? In as much as they are bound to spatial or other representations, they are *constructions*. In as much as they are bound to language, numbering and computing, they are *algorithms*. For an Aristotelian or Leibnizian, mathematics ought to be algorithmic (on the algebraic side) and constructive (on the geometric side). This alone places its logical design under the control of real reason.

Standing against the principle of maximum audaciousness as found in Plato, this approach entails logical prudence and abiding by a set of monitoring procedures.

- Systematic doubt regarding the use of the actual infinite, whether it has to do with the infinitely large or small. This is because the infinite is broadly subtracted from constructive and algorithmic checks and balances. The infinite is *decided*. If the mathematical infinite is conceded to "exist"—no matter what the status of this existence might be— the risk of resuming with Being and forgetting that mathematics is only a logic of possibilities will be great. Even a creature of Leibniz's caliber regarding the field of the differential calculus and integrals intended to keep the real infinite for metaphysics and the divine absolute, which alone would impart a "monadic" status to it. The infinite cannot be defended as numerical or even geometrical: "It belongs to the essence of number, line and some such whole to be bounded." And the "true infinite, strictly speaking, is only in the absolute, which precedes all composition and is not formed by the addition of parts."[1]

- Restricting and monitoring existential assertions in mathematics. The logical essence of mathematics is transparent so long as one remains within the formal consecutions and definitions of possibilities. It is made obscure once 'existence' is announced. One should thus let every assertion of this kind be accompanied by the explicit construction validating it and the logical proof for the case of its existence.

- Tendency to perspectival pluralism. If mathematics is a "formal science," appropriate to the coherent description of possibilities, there is no sense in letting it be unique (as when it is bound to Being and capable of truth). Coexistence of 'different' mathematics can be envisaged. For example, when God's intellect with Leibniz is characterized by the coexistence of possible worlds. They are certainly contradictory amongst themselves, yet remain internally coherent.

The major tendencies of the Aristotelian and Leibnizian grasping of mathematics can be summed up as follows: logicism, algorithmic or constructivist finitism, and a pluralism of rational possibilities.

The ancient Greeks started it all by setting up a constitutive polemic regarding the way philosophy was to seize the mathematical condition. Plato or Aristotle (but also Descartes and Leibniz) lent their names to this discord.

This is a central and complex discord for philosophical thought. On the one hand, mathematics seized by philosophy is bound to the question of Being. This occurs once thought stops struggling with the opacity of experience, but is visibly freed from the constraints of finitude. On the other hand, it is certain that mathematics is paradigmatic as far as rational linkage, consecutions and proofs are concerned. In a broader sense, its logical value is eminent. With respect to construction of philosophical space, mathematics is distributed according to the double register of decision regarding the thought of Being and the formal consistency of arguments. For the philosopher, mathematics is simultaneously ontological and logical. Let us say it is "onto-logical," with the hyphen separating Plato from Aristotle. In my own language, I would say mathematics enlightens philosophy regarding how every truth has an interventional dimension (the axioms, principles, and audaciousness). It also does regarding a dimension of fidelity (formal operators, the continuity of thought, definitions, and prudence). To take up again this double conditioning in relation to my own work (namely, in my proposal to forge a modern concept of truth and do so by betting again on philosophy) is a task requiring a meticulous facing off with mathematical vitality itself.

To make this situation clearer, we most certainly have to take on the great corpus of contemporary mathematics. We must deal with the apparatuses that claim to endow mathematics with its own unified space, or primordial language.

Nowadays there are only two such apparatuses. Each of them was born from the inner needs of living mathematics and not from the outer application of a linguistic philosophy:

- Set Theory, from Cantor to Cohen, emerged in the nineteenth century from the needs of real analysis and topology;

- Category and topos theory, emerged fifty years ago from the needs of geometric algebra.

With respect to the great opposition between Platonism and Aristotelianism, we have to refer to these two apparatuses in order to examine the ontological configurations (or logic of the ontological) that can reanimate the philosophical project in its singularity. What we want to avoid is to concede ground to a specialized domain called "philosophy of mathematics."

Prior to taking on this major crossing (it can only be outlined as a program in this present book,[2]) we have to return to the logical characterization of mathematics, and more generally to the following problem: if mathematics "carts" along with itself logical prescriptions; if, then, its philosophical identification as a science of the pure multiple or first ontology has to be doubled with an identification as onto-logical (in which case Aristotle was not entirely wrong), what are the consequences for philosophy itself? Generally speaking, what is, or ought to be, the relationship between logic and philosophy?

8

LOGIC, PHILOSOPHY, "LINGUISTIC TURN"

The proper mode according to which a particular philosophy summons an experience in thought to its conceptual space does not so much arise from the object's supposed law as from the objectives and operators of that philosophy itself. What is at stake here is not so much whether philosophy ought to be interested in logic, which is now entirely mathematized, due to the fact that logic is a constituted object or given knowledge domain. What we demand instead is to immanently seize this imperative. What we seek to apprehend is how this contemporary seizure of logic is immanent to philosophy.

The properly philosophical take on logic is that the mathematization of logic by Boole, Frege, Russell, Hilbert, Gödel, and many others is narrowly linked to what is called philosophy's linguistic turn. Let us assume a philosophical project to reexamine this linguistic turn; or to identify thought and truths as processes whose language is but one datum among others, or to seek to abandon all transcendental conceptions of language. With any of these research programs, philosophically reconsidering the mathematization of logic is inevitable.

Let me put this more bluntly. If the knot between thought and Being, which is philosophically referred to as "truth," is not of a grammatical essence, or if it falls under the condition of the event, chance, decision, and a-topical fidelity, and not under the condition of anthropological rules and logics of language or culture, then we can ask ourselves what exactly is the ontological determination of mathematized logic.

In my own thinking, this question is quite complex. I would give it the figure of a torsion. Since I posit that ontology, namely, what of Being *qua* Being is inscribed or written as *logos*, is exactly mathematics itself, the upshot is that asking what the ontological determination of mathematical logic is becomes: what is the mathematical determination of mathematized logic?

How is this question philosophical? It appears to refer to a simple internal gap within mathematics. This is the distance from which the status of logic as a mathematical discipline is thought within mathematics itself. Is the thought of this internal gap related to philosophy?

We find ourselves at a complex triangle. It has three poles: mathematics, logic, and philosophy.

In my view, the discriminating axiom to be introduced is the following. Nowadays, a philosophy is broadly decided upon through the position in the relationship it holds with respect to the two other summits of the triangle, namely, mathematics and logic.

The linguistic turn of contemporary philosophy is ultimately marshaled to a large extent by a more or less explicit thesis equating logic with mathematics. Russell's logicism is but an extreme and unnecessary form of this thesis. It is obviously facilitated by the complete mathematization of logic. As I said before, it is a thesis whose provenance is Aristotelian or Leibnizian.

The linguistic turn is known to have two apparently opposing sides. The names that dominate here are Wittgenstein and Heidegger. What we must retain from Wittgenstein is the idea stating a strict coextensivity between the world and language, or that the limits of the latter are exactly the limits of the former. What we must recall from Heidegger is that, in times of distress, thought is foremost on the way to speech. As he wrote about Rilke, "there is veiling because the essential region draws away, but the song naming the Earth remains."[1] In either case, the place wherein the destiny of thought is played out is precisely the frontier of what can be said. For it to be the place, mathematics reduced as it is to computing and blind logic cannot be a thought.

Wittgenstein simultaneously contends that:

1. "Mathematics is a logical method." (*Tractatus*, 6.2), and what we have already cited:

2. "The mathematical proposition expresses no thought" (*Tractatus*, 6.21).

In a single gesture, Heidegger draws mathematics over to the calculation of technical mastery: "This is how the being of beings becomes thinkable in the pure thought of mathematics. Being thus calculable, and placed into calculation, makes Being an entity that can be mastered at the very heart of modern technology. The structure of the latter is mathematical."

Thus, Wittgenstein and Heidegger share the equating of mathematics with logic lying at the heart of a calculational disposition in which thought no longer thinks. They each turn that identification into an appeal to the poem as what persists in naming what is withdrawn. With Heidegger, all that remains for us is the song naming the Earth. Yet Wittgenstein also writes, "I think I summed up my attitude to philosophy when I said: philosophy ought really to be written only as a poetic composition."[2]

The linguistic turn is the philosophically instituted essential correlation between two things. On the one hand, there is the computational equating of mathematics and logic, which is subtracted from thought and slid over to the benefit of a blind and technical power of the rule. On the other, one finds the archi-aesthetic appeal to the pacific and enlightened power of the poem.

Consequently, the prescription to break with this philosophical disposition requires at least two gestures.

First, the critical reexamination of the poem as the support of an archi-aesthetic conception of philosophy's destiny. This is what I undertook in numerous studious on Mallarmé, Rimbaud, Beckett, and Hölderlin. In those essays, I was able to draw out a general philosophical category, the "Age of the Poets." I identified a series of singular operations that drive the notion of poetry as thought (disobjectivation, disorientation, interruption, and isolation). I showed how these operations were inapt to sustain the purpose of archi-aesthetics. But I shall not discuss these points here.[3]

The second gesture reconsiders the association of logic and mathematics to the point of dissociating them. It is capable of restoring a dimension of thought to mathematics, and is undertaken from the standpoint of the thesis according to which mathematics is the thought of Being *qua* Being.

As we have seen, restoring mathematics in this way to its thinking essence leaves off from the idea that Being is the unfolding of pure manifold. As such, mathematics is the science of Being *qua* Being.

The reconsidered dissociation of logic and mathematics arises from a distinction between an ontological decision whose nature is prescriptive and a logical inspection of a descriptive nature. I should like to argue in favor of this point.

What is the method? As for philosophy, I believe it is still subject to the condition of events in thought which are external to it. These events are neither its matter, for philosophy is not a form, nor its objects, for philosophy is not reflexive. They are properly its conditions, that is, what authorizes there to be philosophy or transformation in philosophy in the first place.

Thus, the linguistic turn itself was under the condition of a fundamental event in thought, the mathematization of logic. For, let us not forget, logic used to be that from which philosophy, in its aim to be a thinking appropriation of Being, would seize language for itself. From Aristotle to Hegel, logic was the philosophical category of ontology's dominion over language. The mathematization of logic has authorized language to become that which seizes philosophy for itself. The cost of this was the destitution of ontology itself: either in the shape given to it by Wittgenstein, namely, statements of ontology are non-sense; or in the version Heidegger gives to it, namely, statements of metaphysics are in the epoch of their nihilistic closure.

Let us then ask what event in thought touching upon logic authorizes philosophy to dislodge itself from the dominion of grammar and linguistics? How will we ensure a new internal distance between mathematical thought as such and mathematized logic?

This event can be perfectly identified even if it is still philosophically silent, as Nietzsche used to say in the discussion between Zarathustra and the fire-dog: "The greatest events—they are not our noisiest but our stillest hours."[4] This silent event marks a fundamental change in style for the mathematical presentation of logic. It is about the presentation of logic in the framework of category theory with its core concept of *topos* or "universe." This event began in the 1940s with S. Eilenberg and S. MacLane's creation of a categorical language for the needs of algebraic geometry.[5] It was pursued in the 1950s with Alexander Grothendieck's invention of the concept of universe.[6] It was completed in the 60s and 70s when P. Freyd and F. W. Lawvere reformulated logic as a whole into the language of categories.[7] The concept of elementary *topos* thus became a transparent tool.

Jean-Toussaint Desanti was the first to bring my attention to the fact that an ontology exclusively founded on Set Theory—what he called an "intrinsic ontology"—fell short of recognizing what, in his view, was the capital contribution of a mathematical conception buttressed by the sole datum of morphisms, that is, the regulated correlation between structures.[8]

By placing philosophy under the condition of *topos* theory, I was finally able to solve my problem at least partially and after a very long time of wandering or abstention.

Let me reformulate the problem in a fully articulated way. It will take six theses.

First thesis: We have to break with the linguistic turn that has seized philosophy.

Second thesis: Such a break is required because this orientation in thought ends up with the pure and simple dislocation of philosophical desire as such. Either we attain the Anglo-Saxon space in which philosophy has become a vast scholasticism, a grammar of positions, indeed a pragmatism of cultures. Or we wind up with the Heideggerian dependency, in which one has to entrust the salvation of thought to post-philosophical operations and a fragmentary archi-aesthetics.

Third thesis: At the heart of the conditions of the linguistic turn, there is the formal identification of logic with mathematics, authorized in the final analysis by the mathematization of logic.

Fourth thesis: A new thought has to philosophically be produced on the delimitation between mathematics and logic, albeit acknowledging that logic has indeed been mathematized.

Fifth thesis: Mathematics is posited as the science of Being *qua* Being, or ontology strictly speaking.

Sixth thesis: That logic is mathematized indicates a correlation that has yet to be thought between an ontological decision and a logical form. It is this correlation we seek to bring about in the shape of an irreducible gap.

At this point in the venture, we can shed some light on the difficulty of the problem as well as on what sets the event conditions of *topos* theory for it to be solved.

First the hard stuff. If the mathematization of logic authorized the linguistic turn of philosophy, it is obviously in virtue of the presentation of logic as a syntactical mathematization. By that I mean that its whole purpose, as early as with Frege's ideography, was to constitute logical languages as formal "objects." Under these conditions how can anyone expect a new isolation of logic as such to loosen the grip of grammaticality over philosophy? Even the separation between logic and mathematics can allow linguistic terrorism to subsist in the shape of a pragmatic, cultural, and relativist terrorism, if in each case it drives the mathematician back to the sphere of language and syntax.

You can say, for example, that a formalized theory is logical if its statements are valid in every nonempty model. A formalized theory is mathematical if only one single family of models is appropriate for it. But this delimitation is philosophically inoperative. For the mathematical only appears there as a case of the logical. Or the logical is there in some sense as the universal syntactical substructure of the mathematical. As the appropriate notions of syntax and semantics are a determining factor in the delimitation, the latter cannot release philosophical desire from the dominion of linguistics.

As a variant, you can also say that the mathematical, and not just the logical, is a formalized theory accepting existential axioms that are nonreducible to universal axioms, and then decide upon existence and ascertain its consistency solely regarding this decision. Thus, Set Theory itself would be mathematical in as much as it decides axiomatically on the existence of an empty set and at least on one infinite set. But there again, the delimitation is created by assuming the syntactical common being of logic and mathematics, since their gap concerns the sole differentiating action, as it were, of the quantifiers.

Truth be told, once logic is mathematized in the shape of a syntax or a formal theory, its linguistic connection is primordial. Besides, it is announced symptomatically as such by the field of its designations in natural language: formal languages, formative rules, statements, propositions, syntax, semantics, punctuation, interpretation, and so forth. Thereupon, even the thesis according to which mathematics *is* ontology loses a portion of its constitutive powers. As logic is laid out as a formal language of that ontology, it reintroduces linguistic prescriptions without the ontological decision endowed with the means to revert them.

What is the value then of the event that remathematizes logic now in the framework of category theory? It has to do with a complete shift in perspective. Whereas the syntactical presentation of logic as a formal language lays out universes or models as semantic interpretations, in category-theoretic presentation, what there is are universes. Their logic is *an internal dimension*. In other words, in the linguistic presentation, an ontological disposition is the appropriate referent of a formal theory. This is obviously what authorizes the infinite Anglo-Saxon gloss separating and articulating the formal from the empirical. In the categorical presentation, one proceeds to geometric descriptions of universes and notices that to such and such a disposition of universes such and such a logical disposition is immanently correlated. Therefore, logic becomes an intrinsic dimension of possible universes. Or, more fundamentally, the descriptive characterization of a thinkable ontological state entails logical properties that are themselves presented in the space of Being, or the universe, that thought describes.

In this reversal, two things disappear:

- First, the formal and linguistic antecedence of the logical, or more generally the grammatical, over the position of a universe or over an ontological decision;

- Then, the way mathematics is enfolded by logic. In fact, the logical appears as an immanent constraint enfolded by the mathematical. And especially, the logical is localized. It is a presented and localizable dimension of universes. Mathematics takes on the task of describing their possibility.

The problem of setting up a delimitation between mathematics and logic then takes on a completely other twist. This delimitation no longer allows itself to be decided upon by the linguistic criteria that used to deplete its strength. Instead, it is sent back to much more fundamental distinctions, which are themselves ontological. These distinctions concern two conceptual couples: the real and the possible, and the global and the local. This is what can be called an essential ontological "geometricization" of the relation and "de-relation" between logic and mathematics.

9

FIRST REMARKS ON THE
CONCEPT OF *TOPOS*

If it is true that *topos* theory internally ascertains logic as a local dimension of possible mathematical universes, there must exist theorems in this theory which I would call "ontologico-logical." We could also shift the hyphen and call them "onto-logical." Let some theorems have the following generic form: if a thinkable universe possesses some such ontological characteristic, then some such logical constraint can be spotted in it. These theorems accomplish *both* a reversal of perspective *and* withdraw any explicit mention to language and its regulated syntactical usages.

Theorems of this type in fact do exist. I would like to share my feeling now regarding their eventful strength for philosophy. I am going to discuss two such theorems, though I do intend to keep my discussion within the bounds of natural language. They are linked to the ontological concepts of difference and the void (or of nonbeing), respectively.

A. Starting with difference, the essential philosophical problem we encountered with respect to the characteristics of a manifold ontology and in opposition to Deleuze's "intensive" ontology has to do with intelligibility. Is some such nondescript difference always locally recognized? I mean is every difference thinkable at a single point in the differentiating action that unfolds it? Or do qualitative differences exist which are thinkable only globally? This question clearly opposes ontologies that can be said to be discrete, like with the atomists, the first Wittgenstein or even Plato, to continuist ontologies like Bergson or Deleuze's. The key to thinking through this problem lies with Leibniz. Leibniz is the one who tried to integrate absolute local differences, that is, the "metaphysical points" embodied by monads, to a global or integral difference of the universe.

The notion of difference is mathematically representable in the description of some such topos, and so are notions of localizable difference, or difference recognized in a point. A topos in which every difference is localizable in a point is said to be "well pointed." It is clearly an ontological feature of a topos to be well pointed.

As I said, its internal logic can be identified. The fact that this is well known stems largely from the immanent and local data of a topos. To give a couple of elementary examples, the true and false are local data of a *topos*, that is, "simple actions" that are part of the universe. By the same token, negation, conjunction, or even implication is also part of the universe. All of these terms designate relations that have been identified, and are present "in person" in the topos. There are no syntactical preambles or semantic interpretations.

A topos can also have its immanent logic characterized. For example, it can be classical or nonclassical. In short, a classical logic validates the law of the excluded middle. But the definition of classicism in a topos is given in terms of its construction. The way this is said is as follows. Incidentally, this will be my only technical citation so as just to murmur the sound of it: the logic of a topos is classical if the sum of the true and the false is an isomorphism. What interests us is the proof of the following theorem: "If a topos is well pointed, then it is classical."

This is a remarkable onto-logical theorem. If the universe, or space-being, accepts such a property regarding the intelligibility of difference (in this peculiar circumstance, that it always be locally recognized), then its logic is necessarily classical. Accordingly, it can be shown that an ontological singularity of the presentation of difference entails a purely logical constraint. We slide from the manifestation of Being to the principles of language, and not conversely.

B. My second example involves the void as well as Zero. As a way of numerating absolutely noncomposed multiplicity, the void is a major ontological plus. It is the matheme of the suture of any discourse to the Being by which it is sustained. Regarding the void, the key question is that of its unity. If ontological description fixes the unity of the void, in the Parmenidean tradition it assumes a certain reversibility of Being as a subtraction from counting and the One. If description accepts the multiplicity of the void, or its absence, then it pluralizes its foundation. In the Heraclitean tradition, it is instituted as alteration or becoming.

It is easy to make sense in a topos of the notion of empty object or empty letter. The following astonishing theorem may then be proved, as well as shown to be an "onto-logical-onto" theorem. Let a constraining correlation between the coupling of an ontological and logical feature on the one hand, and another ontological feature, on the other. The theorem then reads as follows: If a topos accepts only a single empty object, and if its logic is classical, then it is a well-pointed topos. What this means is that if you are in the reciprocity of Being and the One, and your logic is classical, then your ontology of difference accepts that every difference is recognized in a point, and that there is no purely qualitative or global difference. This time, a complex dimension of the possible universe, that is, an admixture of Being (the void is one) and principle (the excluded middle), constrains a simple ontological feature, which is the status of Difference.

These two examples show the subtle networks woven by the categorical presentation of logic. They stretch between its primordial ontological determinations, like difference and the void (but there are theorems to cite on the infinite, existence, or relation which are no less remarkable), and its immanent logical determinations, like the validity of the excluded middle. These networks may pertain to classical logic, as they may also to intuitionistic, constructivist or nonclassical imperatives. For example, if a topos is not well pointed, and hence if there exists as with Leibniz, Bergson, or Deleuze, intensive, qualitative, or global differences, and if in addition you accept a certain reciprocity between Being and the One, as Leibniz does, then it is impossible for your logic to be classical. This is a reciprocal upshot resulting from my second example. Strictly speaking, if one takes account of the insensible degrees of Being, Leibniz's logic is not really classical. For we are not absolutely constrained to decide between states p and not-p, since between them infinitely many intermediary states exist.

10

FIRST PROVISIONAL THESES ON LOGIC

Let us return to our initial problem. Under condition of topos theory how do we go about rethinking the gap between logic and mathematics? How then do we start restituting mathematics' dimension of thought, that is, its Platonic dimension? Finally, how to interrupt the linguistic dominion over philosophy's destiny?

What *topos* theory offers is a description of possible mathematical universes. Its method employs definitions and schemas, and a geometric synopsis of its resources. It is tantamount to an inspection of Leibniz's God: a categorical journey through thinkable worlds, their kinds and distinctive features. It ascertains that each universe bear its own internal logic. The theory establishes the general correlations between ontological features of these universes and the characterization of their logic. But it does not decide on a particular universe. Unlike Leibniz's God, we do not have any reason to consider some such mathematical universe as the best of possible universes.

From this point of view, and in virtue of its being a rational inspection of possibilities, topos theory is appropriate to logical thought. It is the apparatus wherein the logic of the constraining correlations between ontology and logic is thought out. It could be defined as the mathematized logic of the "onto-logical."

Now ontology is not reducible to the logic of the onto-logical. Real mathematics is not a mathematized inspection of possible mathematical universes. Real mathematics decides upon one universe. The upshot is that the relation between logic and mathematics is like the one existing between a general investigation of an ontology's logical resources and an ontological decision that entails its logical consequences. This also means that logic is definitional, whereas real mathematics is axiomatic. This is how, if you axiomatically decide

119

that the thought of Being is crafted in the framework of Set Theory, you will have to accept that your logic is classical. Why? Simply because when considered a topos, and hence examined logically, Set Theory is like a well-pointed topos at which any difference is recognized in a point. As we have seen, the upshot of this is that the immanent logic of a well-pointed topos has to be classical. Remaining with the Leibnizian imagery, logic will then be said to think possible worlds, and ontology to decide upon one world, in the way God "fulgurates" the monads.[1] The major difference here is that no calculation external to a decision may propose a law of the best.

So let us lay out the thinking space in which we break with philosophy's linguistic turn. This can be articulated in a short series of theses:

1. Logic is not a formalization, syntax or linguistic device. It is a mathematical description of possible mathematical universes beneath the generic concept of topos. A mathematical universe, a topos, localizes its own logic.

2. A possible mathematical universe lays out its constraining correlates between certain ontological features and certain features of its immanent logic. The study of these correlations comprises the fundamental content of logic itself. In short, logic thinks its own subordination to ontology. Insofar as it thinks this subordination, it can be mathematized, since mathematics is ontology itself.

3. Mathematics is accomplished by axiomatic decisions that set up a possible universe as real. The outcome incurs logical constraints. The latter are logically thought out by the logic of possible universes. They are practiced but not thought by real mathematics.

4. Consequently, the irreducible gap between logic and mathematics holds by a blind spot to a thought-decision, which is that every decision of this kind installs a logic that is practiced as necessary, although it is a consequence of the decision. Mathematical logic is a clearing of this blindness, for it thinks the onto-logical correlation. Yet to do so, it must regress from the Real to the Possible, as the Real is only encountered beneath the axiomatic imperative. The

Possible only lends itself to description provided it occurs through various regimens of definitions and classifications.

The paradox arising from this series of theses is that the word "logic" shows a double occurrence. "Logic" is called at one and the same time what is localized as such in a possible universe with its singular figures: the true, the false, negation, quantifiers, and so forth. But it is also called the mathematized thought of constraining correlations between the ontology of the universe (difference, void, infinite, etc.,) and this logical localization.

This paradox is merely the paradox of any universe. Its purely logical determination, arising from definitions and the possible, embodies a coming-and-going between the local and the global. There will be a local definition of logic: a particular machinery localizable in the *topos* and technically articulated through the notion of "subobject classifier." Then a global definition will be used to think through the systematic correlations between the being and action space of *topos* and its logical singularity. It is this relation between the global and local which ends up rendering a geometric dimension to logic.

This is why a fifth thesis has to be added:

5. Delivered from its syntactical and linguistic hold, logic is always geometric logic. In other words, it is also the (global) logic of (local) logic.

This thesis is compatible with Wittgenstein and Lacan's arguments on the inexistence of any metalanguage because logic as a thought is here unbound from formal language. As a dimension of a possible universe, any logic—whether it be a logic of logic or a logic of the onto-logic—stands without presupposing a metalanguage.

The consequences of all this for philosophy are considerable. They clearly lay out a program for thought:

• We need to prompt a break with the linguistic hold in its dual form: Anglo-Saxon analytic philosophy, ordinary language philosophy and pragmatism on the one hand; the hermeneutic of the Dasein and the "archi-aesthetic" of the poem, on the other. This double break is conditioned by an event, which attests to its contemporaneity.

- We need to confront this rupture with the philosophical traditions of the century that are not inscribed in the linguistic turn. I am thinking particularly of all that stems from Nietzsche and Bergson, and is articulated in Deleuze's thought. Deleuze's imprecations against logic are well known. We have said enough for it to be understood that the *disputatio* focuses not on ontology per se, but the status of multiplicities, the thought of difference and thus on what is thought beneath the anonymous name of Being.

- Now we have to reexamine logic from the perspective of ontological decision itself. Hence, we need to think the possible from the basis of the Real. This also means we have to invest definitions from the standpoint of axioms, and not vice versa. Finally, we have to posit that, as a thought apparatus, philosophy is essentially axiomatic—and not definitional or descriptive.

- Thereof we have to restore and reshuffle the category of truth. We must show that the existence of a logic depends upon a truth process that is itself conditioned by a chance event as well as a decision about that chance event. After which the task is to show that there is a logic of truth, but not a truth of logic.

- By this very fact, we need to ascertain that there is something by nature left unthought in all truth. Working upon this point retroactively, as it were, a new thought of sense has to be elaborated.

As often happens, many points of this program were anticipated by the poets of the Age of the Poets and outstandingly so by Stéphane Mallarmé.

Mallarmé really saw that a decision in thought, that is, for him a decision of the poem, which is an encounter or trial with chance, induces what he called the "motif of a logic." This notion of logic is really the logic of the onto-logic owing to how it lays out infinity. The latter cannot be prior, transcendental, or linguistic, because it is suspended to an event, or to an Act. This Act takes the logically inverted shape of an encounter with chance, in the guise of throwing dice. Therefore, thought is under condition of an adjunction of a pure event. For this reason it produces an implacable logic. The latter arises from the fact that thought is exposed by accepting the Act and by being faithful to the event. In turn, this fidelity lays out a truth that is not necessary for anybody, save for its Subject. Necessity is but a result.

A passage from Part I of the fragments of *Igitur* recapitulates that to which philosophy is still summoned today:

> But the Act accomplishes itself.
>
> Then his self is manifested in his reassuming Madness, admitting the act, and voluntarily reassuming the Idea as Idea, and the Act (whatever the power that guided it) having denied chance, he concludes from it that the Idea has been necessary.
>
> Then he conceives that there is, to be sure, madness in admitting it absolutely: but at the same time he can say that since through this madness, chance was denied, this madness was necessary. For what? (No one knows, he is isolated from humanity.)
>
> All there is to it is that his race has been pure: that it took from the Absolute its purity to be so, and to leave of it only an Idea itself ending up in Necessity; and that as for the Act, it is perfectly absurd except as movement (personal) returned to the Infinite: but that the Infinite is at last *fixed*.[2]

With Mallarmé, that the infinite be finally "fixed" means it dons the form of the number of what *Dice Thrown Will Never Annul Chance* designates as "the sole number that cannot be another." This correlation of number and fixed-Being is often repeated, but nowhere with as much brilliance as in the sonnet in *ix* and *or*. Let us gaze at its denouement:

> She, dead naked in the mirror, in body again (*encor*)
> That in the oblivion enclosed by the frame is fixed (*fixe*)
> With scintillations at once the septet (*septuor*).[3]

What ends up set as the emblem of the 'pure notion' wherein Being is captive, is in fact the stellar integer, the seven stars of the Big Dipper.

Yet can the infinite be an integer? This is what Mallarmé, Cantor's unconscious contemporary, contends in the poem. That the infinite is a number is what a set-theoretic ontology of the manifold finally made possible after centuries of denial and enclosure of the infinite within theology's vocation.

This is why the ontology of number is an important item in the secularization of the infinite. Indeed, it is the only way to be freed from both religion and the romantic motif of finitude.

11

THE BEING OF NUMBER

As the Euclidean definitions indicate, the Greek thought of number suspends its being to the metaphysical aporiae of the One. In definition 2 Book 7 of Euclid's *Elements*, the number is "a multiplicity comprised of unities." And unity, in definition 1 of the same book, is "the basis from which 'one' is stated for every being." Finally, the being of number is the multiple reduced to the pure combinatorial legislation of the One.

The saturation and collapse of this thought on the being of number as the One's procession signs the entry of the thought of Being into modern times. Its collapse combines three factors: the Arab appearance of zero, calculus of the infinite and crisis in the One's metaphysical ideality. The first factor, zero, introduces neutrality and the void into the heart of the thought of number. The second factor, the infinite, draws combinatorial calculation towards topology, or adjoins the numerical position of a limit to mere succession. And the third factor, the One's obsolescence, constrains us to think of number directly as a pure manifold, as a manifold without-one.

The upshot of this is a kind of anarchic dissemination of the concept of number. It can be seen in the disciplinary syntagma of "number theory," which proves to contain huge chunks of pure algebra and particularly refined panels of complex analysis. We also see this in the heterogeneity of introductory procedures of different classical number types: axiomatic for natural numbers, structural for ordinals, algebraic for negative as well as rational numbers, topological for real numbers, and mainly geometric for complex numbers. And, finally, this is seen in the non-categoricity of formal systems attempting to capture number. These systems accept nonclassical models, open the rich avenues of nonstandard analysis and consequently restore all rights to infinite or infinitesimal numbers.

For a philosophy whose purpose is to show that there is an active thought of Being in numbers, what is most difficult is that apparently—and this is quite unlike old Greek times—there is no single unified definition of number. Which concept can simultaneously subordinate the discretion of natural numbers, density of rational numbers, completeness of real numbers, swarm of infinitesimals, let alone the transfinite numbering of Cantor's ordinals? How can all this arise for philosophy from a concept having to endure and magnify its power of thought in the same stroke as its inventive singularity? Let us try to clear up this disorder by beginning with the ordinary functions of the word "number."

What do we mean by 'number' if we consider the linguistic and representational usages to which it is linked?

First of all, that a number is an instance of measuring. Basically, a number serves to discriminate the more from the less, and the greater from the smaller. It calibrates data. A kind of number will then be required to give us a structure of order.

Second, a number is a figure for calculating. One counts with numbers. Counting means adding, subtracting, multiplying, and dividing. A number kind will be required to have practicable or well-defined operations. Technically this means that a number kind ought to be identifiable algebraically. The completed recapitulation of this identification is the structure of an algebraic group wherein all operations are possible.

Third, number ought to be a figure of consistency. This means that two determinations, that is, order and calculation, have to obey rules of compatibility. For example, the addition of two clearly positive numbers is expected to be greater than each of these numbers. Or the division of a positive number by a number greater than the unit is expected to yield a smaller result than the first number. These are the linguistic requirements of the idea of number, matched by order and calculation. Technically, it is said as follows: the adequate figure in which is a number kind is inscribed is that of the ordered group.

Thereupon, if a definition of number has to subsume each of its specific kinds, this means that it has to determine what I call the "ordered macro-group" in which all number kinds lend themselves to a localization.

This is precisely the result of the definition the great mathematician John Horton Conway proposes beneath the paradoxical name of "surreal numbers."

In the general framework of Set Theory, this definition specifies a configuration wherein a total order is defined, and addition, subtraction, multiplication, and division are universally possible. We then observe that in this configuration or in the macro-group of numbers, ordinal and natural numbers, the ring of positive and negative integers, the rational and real groups, and all of their known structural determinations are found. But we also observe infinitely many yet to be named number kinds, infinitesimals in particular, or numbers situated "between" two adjacent and disjointed classes of real numbers, or infinite numbers of all kinds apart from ordinals and cardinals. That we are actually dealing with a macro-group is borne out by the fact that the latter has nothing to do with sets. This is why I have called it a "configuration." It is about a class proper. Moreover, this should be obvious given that it contains all of the ordinals, which already do not constitute a set. Joining up with an intrinsic determination of manifold-being as such, what designates the concept of number is an inconsistent multiplicity. Number kinds cut out consistent numerical situations in this inconsistency, which is the content of their being. This is the way the real numbers group is consistent and a set. But its identification as a group of numbers refers to the idea that there is an internal consistency to the inconsistency of the number locus, or that there is a subgroup of the numerical macro-group.

The apparent anarchy or conceptless multiplicity of number kinds can be said to be an outcome of how hitherto they were performed in their operations, but not situated in their being. With the macro-group, we have the inconsistent generic locus wherein numerical consistencies coexist. Thereupon, it is legitimate to consider these consistencies as partaking of a single concept, namely, Number.

The being of Number as such, that is, what it is in a number per se which "thinks" Being, is finally given in the definition of the macro-group as an inconsistent space of the being of the consistency of numbers.

We call "Number" every entity belonging to the macro-group. And we speak of "numbers" when designating the diversity of kinds, or the immanent consistencies whose space is laid out by the inconsistency of Number.

What then is the definition of a Number?

It is a definition of admirable simplicity: a Number is a set of two elements, namely, an ordered pair, consisting of an ordinal and a part of this ordinal, respectively. A Number will therefore be written as (α, X) where X is a part of ordinal α, or $X \subseteq \alpha$.[1]

This definition may certainly appear to be circular. It summons the ordinals we have stated as numbers present in the macro-group.

But it is actually possible to define ordinals at first in a purely structural fashion without recourse to any numerical category whatsoever, or, notwithstanding their name, to an idea of order. In von Neumann's definition an ordinal is a transitive set in which all of the elements are transitive. Admittedly, transitivity is an ontological property: it signifies only that all of the elements of the set are also parts of the set. If you have $\alpha \in \beta$, you also have $\alpha \subseteq \beta$. This maximum correlation between membership (or element) and inclusion (or subset, part) provides transitive sets with a peculiar kind of ontological stability that, as far as I'm concerned, designates its *natural* being. It is this natural stability of ordinals, this immanent homogeneity, which designates them as the primordial stuff of Number.

Moreover, what is striking in the definition of Number—the ordered pair of an ordinal and a part of this ordinal—is the instance of the pair. The task of defining Number obliges us to settle into the two. Number is not a simple mark. There is an essential duplicity to it. But why this duplicity?

To borrow from Gilles Châtelet's vocabulary, Number is a gesture in Being and the double marking is a trace of this gesture.[2] On the one hand, we have a homogenous and stable mark, the ordinal. On the other, we find a mark that is somewhat torn from the first. It is an indeterminate part that most often fails to conserve any immanent stability. It can be discontinuous, dismembered, or without any concept whatsoever, because nothing is more errant than the general idea of the part or subset of a set.

The numeric gesture is the somewhat forced, deregulated, and inventive withdrawal of an incalculable portion of what has all of the attributes of order and internal solidity.

This is why as a philosopher I have renamed these two components of a Number. I call the ordinal, "Number's *matter*," in a bid to evoke the donation of stability and an internal architecture that is potent, albeit almost indifferent. In turn, I have called the "part" of the ordinal, "Number's *form*." This is not meant to evoke its harmony or essence. Instead, it designates, as in some of contemporary art's effects, what is inventively wrenched out from a still readable background of matter. It is what draws an unpredictable, almost devastating discontinuity from matter, which allows an inalterable material compactness to appear as if through its lacunae.

Thus, a Number is entirely determined by the coupling of an ordinal matter and a form cut out from this matter. It is the duplicity of a compact figure of manifold-being, and of a cut-out gesture devoid of rules in this compactness.

It is remarkable that, from the basis of these ever so simple data, one can ascertain all of the properties of order and calculation we might expect from a way of locating the correlate of the word 'Number' in Being itself.

Now we come to the technical part of things. It can be shown that the universe of Numbers is totally ordered and a group structure (namely, adding, multiplying, subtracting, and dividing) can be defined in it. This is how the construction of an ordered macro-group as a space of ontological identification of everything falling beneath the concept of Number is achieved.

Thereafter, we ascertain that the usual number kinds are consistencies that have been cut out of this space. Natural, relative, rational, and real numbers are subkinds of the macro-group. They are numbers identifiable within the ontological space of Number.

These historical examples aside, many other entities, strange, unidentified or unnamable entities, arise massively from beneath the concept of Number.

Let us give two examples:

1. We are used to considering negative finite numbers. The idea of a negative of the infinite is no doubt stranger. However, in the Number macro-group, the negative of an ordinal, whether finite or infinite, is easily defined.

2. In the macro-group identifying the place of Number, real numbers are shown to include all of the Numbers whose matter is the first infinite ordinal, that is, ω, and whose form is infinite. What about Numbers whose matter is an infinite ordinal greater than ω? Well, we can say that they

are generally about Numbers that have not yet been studied or named. This is a reserve of infinitely many infinite kinds of Number destined to future open-ended investigations concerning the Being-figures of numericity. This entails that the numbers we use on a practical basis are de facto but a minute portion of what resides in Being beneath the Number concept. This can also be said as follows: the ontological prescription held by the concept of Number infinitely exceeds the effective historical determination of named and known numerical consistencies. There is more being beneath the word 'Number' than what mathematics has been hitherto able to circumscribe and catch within the nets of its constructed consistencies.

In fact, within each of its segments, be they apparent or minute (as far as our intellect is concerned), the macro-group of Numbers is populated by an infinite infinity of Numbers. The best possible image of this universe is given by Leibniz in a description from the *Monadology*, paragraph 67: "Each portion of matter may be conceived as like a garden full of plants and a pond full with fish. But each branch of every plant, each member of every animal, each drop of its liquid parts is also some such garden or pond."[3] Every minute interval of the Number macro-group can be conceived as the space of infinitely many Number kinds. In turn, every kind and every minute interval of that kind comprise yet another such place or infinity.

What to conclude from all this?

Number is neither a conceptual feature (Frege's theory), nor an operatory fiction (Peano's theory). Nor is it an empirical and linguistic datum (pop science theory), nor a constituting or transcendental category (Kronecker or even Kant's theory), and not even a syntax or language-game (Wittgenstein's theory). Finally, it is not even an abstraction of our idea of order. What Number is is a form of manifold-being. More precisely, the numbers we regularly manipulate make up but a minute sampling of Being's infinite prodigality in terms of Number kinds.

Essentially, a Number is a form wrenched from stable and homogenous manifold-matter. The concept of this matter is the ordinal in the intrinsic sense given to it by von Neumann.

Number is not an object or objectivity. It is a gesture in Being. Prior to all types of objectivity, prior to any presentation connected with its being or disconnected from it in eternity, Number opens up to thought as a formal cutout from the maximal stability of the manifold. It is numbered by matching that stability with the most often unpredictable results of the gesture. The name of Number is the duplicitous trace of components in the numerical gesture.

Number is the place of Being *qua* Being as far as the manipulable numericity of number species is concerned. Number '*ek-sists*' from numbers as a latency of their being.

What is remarkable is that we actually have some access to this latency. We have access to Number as such, even though this access points to the excess of Being over knowledge. It is an excess manifest in the innumerable extension of Numbers in contrast to what we know about structuring them in terms of the presentation of number kinds. The fact that mathematics at least allows us to designate the excess and have access to it, confirms this discipline's powerful ontological vocation.

Regarding the concept of Number or any other concept, the history of mathematics is exactly the history of the relation between the inconsistency of manifold-being and what of this inconsistency our finite thinking manages to make consistent. It is a never ending history, whose very principle is endless.

The task pertaining to Number and numbers can only be to pursue and ramify their concept's unfolding. Number is exclusively part of mathematics once the task is to think its kinds and situate them in the macro-group that is their place. Philosophy states how Number belongs exclusively to mathematics. It designates where numbers are given as Being's resource within the limits of a situation—the ontological or mathematical situation.

The thought of Number and the avenues followed by Frege, Peano, Russell as well as Wittgenstein have to be abandoned. What has to be done instead is to take Dedekind and Cantor's undertakings to their radical limits, overwhelming them and thinking each to the point of dissolution.

Number is neither deduced nor induced. At this point language and sensible experience become inoperative guides. The task is only to remain faithful to what emerges from Being's inconsistent excess as a consistent historical trace in the movement by which mathematics is at once interminably and eternally recast, and to which our thought has at times been bound.

One may object to this line of reasoning by putting forward the operative Kantian conception in contrast to the inexhaustible density of the ontological concept of Number. The Kantian conception refers to number as a mere schema of succession. As such, the transcendental form of time, that is, the intimate plinth of sensibility, induces it for all types of constitutive understanding. For Immanuel Kant, the upshot of this was that the idea of a being of number amounts to non-sense.

In the final analysis, though, the root of this opposition resides, as always, in what Kant held to be an ontological decision. That this ontology is subtractive is self-evident once the "thing itself" remains inaccessible to us. That it be impossible or inexistent is something to be inferred from a merely cursory interpretation of Kant. Heidegger for one had a strong intuition of this.

This is why this journey through the thinking of Being will only become clearer after we have crossed the Kantian places of a thought on Being.

12

KANT'S SUBTRACTIVE ONTOLOGY

At first glance if there is no ontology with Kant, if even he declares such a task to be inconsistent, the reason is surely that Kant is the philosopher par excellence of the relation or linkage by which phenomena are connected. This constitutive primacy of the relation prohibits any access to the being of the thing as such. Are the famous 'categories' of experience not a veritable conceptual repertoire of all the linkage types that are thinkable (inherence, causality, community, limitation, totality, etc.)? Is it not a case with Kant of ascertaining that the ultimate foundation of the *linked* nature of representation cannot be found in the being of the represented? That the transcendental subject must be tagged onto it through its constituting and synthetical effect? If so, we could imagine the Kantian solution to the problem of structured representation as consisting in a distribution of the pure inconsistent manifold (that is, in my own ontological conception, Being *qua* Being) on the side of the phenomenality of the phenomenon. As for the count-as-one (namely, in our conception, the being-given, or being 'in situation'), it would be distributed on the side of the relation, whose development would itself be based on the structuring activity of the Subject. Consistency would befall to the experience of the phenomenal manifold by the specific potential of the count-as-one found in the universal links the Subject prescribes to it.

Admittedly, this is not what occurs. In one of his most radical intuitions on experience, Kant firmly distinguished *combination (Verbindung-liaison),* which is the synthesis of the diversity of phenomena, from *unity (Einheit),* which is the original foundation of connection itself: "Combination is representation of the *synthetic* unity of the manifold. The representation of this unity cannot, therefore, arise out of the combination. On the contrary, it is what, by

133

adding itself to the representation of the manifold, first makes possible the concept of combination."[1]

Far from being solved by the relation categories, the problem of inconsistent diversity in the count-as-one must itself be settled in order for the relational synthesis to be possible. Kant very clearly sees that the consistency of manifold-presentation is original, and that the links organizing the phenomena therein are only derived realities of experience. The question of the qualitative unity of experience puts linkage *in its rightful place*, which is second. The fact of experience presenting one-manifolds must be grounded beforehand. It is only after this that the origin of phenomenal combinations may be thought.

Putting it in other terms, we have to understand that the source of *order* in experience (the synthetical unity of the diverse) cannot be the same as that of the *One*. The former is found in the transcendental system of the categories. The latter is necessarily a special function Kant may well have assigned to the understanding but that categorical "functioning" presupposes. This supreme function of the understanding, vouching for the general unity of experience and hence for the "law of the One," is what Kant calls "original apperception." If the subjective connotation of original apperception is left aside, conceived as it was by Kant as the "transcendental unity of self-consciousness," and if we set our sights on the operation per se, what I call counting-as-one would be recognized effortlessly. Kant applies it to the universal and abstract situation of representation in general. What original apperception names is that nothing reaches presentation without being a priori subjected to the determination of its unity: "Synthetic unity of the manifold of intuitions, as generated *a priori,* is thus the ground of the identity of apperception itself, which precedes a priori all *my* determinate thought."[2] What makes combination or linkage possible is not linkage itself. From this perspective, linkage in-exists. Rather the faculty of linkage, which does not refer to effective relations since it is enumerable only from the One, is an original law of the consistency of the multiple. It is the capacity of "bringing the manifold of given representations under the unity of apperception."[3]

The distinction between the counting-as-one, that is, consistency's guarantor and the *original* structure of every presentation, and the linkage characterizing *representable* structures, is therefore indisputably thought by Kant from within the transcendental activity of the understanding. He does this according to the gap between pure original apperception (the function of unity) and the system of categories (the function of synthetical connection).

Yet Kant introduced original apperception merely as a condition for the complete solution to the relation problem per se. It is Kant's elucidation of order, his knowledge correlate, which bids him to thinking the One. What I mean can be stated as follows and was already firmly pointed out by Heidegger: what is always problematic with Kant is not the critical radicality of the conclusions, whence his audaciousness excels. Instead, it is how singularly narrow the access is to this radicality. It is not the possibility of presentation in general that begins his approach. The first question for him is to figure out how synthetical a priori judgments are possible, that is, the universal constants of combination which he believes are found in Euclidean mathematics or in Newton's physics. Although his approach finds its starting point in a probably inexact analysis on the form of scientific statements, its rigor leads him to conditions and radical distinctions—like those of unity and connection. However, the limitative effect of this original point resonates in the consequences, which do not always clearly deliver the real extension of their meaning.

Approaching the "there is" from the basis of "there are connections" does not leave the doctrine of the One untouched. With Kant, there is a clear *trace* that the supreme function of counting-as-one is put forward solely in a need to bolster the connecting activity of the categories with an original consistency. Thereupon this "one" will only be thought *for* the connection. The limit concept of consistency will be based on what the relation of the phenomenally diverse calls for. The fundamental structure of presentation will be ordered according to the illusory structure of representation. The trace subsumes the originality of the presentation of manifold-as-One beneath its necessity for thinking representable connections. Insofar as it does, it resides in the notion that with Kant the manifold-One is borne out in the limitative figure of the object. If in the end Kant is only able to think the manifold-One in the representable straight and narrow of the object, it has to do with how he ends up subordinating, within the discursive movement itself, the object's presentational consistency to the solution of a critical problem conceived as an epistemological problem. As Heidegger so accurately described, Kantian ontology bears the shadow of its initiation into the pure logic of cognizance (*connaissance*).

In real terms, the object is in no way a pertinent category by which to designate the existent, such that it is recognized in a situation as being the counted-one of pure manifold. Object designates the One only with respect to connection. Object is what is representable of the existent according to the illusion of linkage. The word 'object' serves as an illegitimate mediation between two disjointed problems, namely, that of the count-as-one of the inconsistent manifold (namely, the appearance of Being) and that of the linked empirical fact of existents. The notion of object is equivocal when faced with that other typically Kantian equivocation in which a single term, the "understanding," is allotted with two separate functions: the supreme function of unity—original apperception—and the categorical function of connection.

When Kant writes that "the transcendental unity of apperception is the one by which the manifold as it is given in intuition is reunited in a concept of object," he makes the manifold-One fall back on the object in such a way that the same term also serves to designate what the bonds connect in representation. It is correlated with original apperception regarding the way it lays out the One in manifold presentations. Likewise, the object will be just as much laid out for the categories, which are conceived as "concepts of object in general by which the intuition of this object is considered *determined* in respect of one of the *logical functions* of judgment."[4] That what exists in experience is an object ensures the "double entry." It is ontological—according to the One (nonbeing) of Being (manifold)—and epistemological—according to the logical form of judgment. This is how Kant sutures his discourse. But the object's equivocation, apart from intending to ground linkage per se—about which David Hume was right in the end to think that it had no being and that it was a fiction—has an inconvenience, which is to weaken the radical distinction Kant had audaciously reached between the origin of the One and the origin of the relation.

The fact is Kant holds steadfast to his conviction that the a priori conditions of the phenomenal linkage must include, under the name of "object," the supreme condition of the representational field. There is no other sense to his celebrated formula: "The a priori conditions of a possible experience in general are also at the same time conditions of the possibility of objects of experience."[5] The word "object" here works intentionally as a pivot between the condition of the consistency of presentation, thus referring to the manifold as such—to the original structure—and the derived condition of the bond between representable "objects," here referring to the empirical manifold—to illusory situations.

Admittedly, Kant knows full well that the object leaves the "being of the object," its objectivity, the pure "something in general=x" undetermined. The connection to its being is underpinned by the latter, without x ever being presented or connected itself. We know this x is the pure or inconsistent manifold. Insofar as it is the correlate of apparent linkage, the object has no being. Kant is keenly aware of the subtractive character of ontology, of the void by which the presentative situation is conjoined with its being. The existent-correlate of original apperception, conceived of as a nonoperation of counting-as-one, is not exactly the object, but the form of the object in general. It is absolutely undetermined Being, from which the fact that there is an object proceeds. At the height of his ontological meditation, Kant ends up thinking the operation of counting as *the correlation of two voids.*

Kant splits up the two terms of the subject/object couple. The empirical subject exists "according to the determinations of our state in inner perception." It is a changing state with no stability or permanence. Corresponding to it is the represented phenomena that "have, as representations, their object and can in turn be the object of other representations."[6] The transcendental subject is given as such in original apperception, which is the supreme warrant of objective unity (thus, of the unity of object representations). In relation to the latter, "every object representation is only possible" and is "pure, original and immutable consciousness." Corresponding to it is an object "that can no longer be intuited by us" since it is the form of objectivity in general, and is opposed to empirical objects as a "transcendental object=x."[7] This object is not one of "several" objects. It is *the generic concept of consistency for any possible connected objectivity,* that is, the principle of a donation-of-the-One from which there are objects *for* connections. The transcendental object is "in all the knowledge we possess always of the same kind=x."[8]

There is the subject of experience (immediate consciousness of self) and its multiple vis-à-vis, namely, the objects connected in representation. There is original apperception (pure and unique consciousness) and its *vis-à-vis*, the object of objectivity, the presupposed x from which the linked *objects* draw the form of the One.

Admittedly, the common characteristic of original apperception as a proto-transcendental subject and of the x that is the proto-transcendental object is that this subject and this object are absolutely unpresented. They are invariant primitive figures whose possibility is all that representation requires. The subject and that object are designated outside of all experience only as the void withdrawn from Being, about which all we have are the names.

The subject of original apperception is only a necessary "numerical unity." It is an immutable power-of-One and it is not knowable as such. Kant's entire critique of the Cartesian *cogito* is founded upon the impossibility of holding the absolute power of the One of the transcendental subject for a knowledge domain, that is, in order to determine a real point. Original apperception is an exclusively logical form, an empty necessity: "Outside of the logical meaning of the Ego, we have no knowledge of the Subject in itself which, as a substrate, would be at the foundation of the logical subject and all thoughts."[9]

Regarding the transcendental object=x, Kant explicitly declares that it "is nothing for us" as it has to be "something different for each of our representations."[10]

The subtractive radicality of Kantian ontology ends up placing the relation between an empty logical subject and an object that is nothing at the foundation of representation.

This is why I cannot consent to Heidegger's idea on the differences between the first and second editions of the *Critique of Pure Reason*. For Martin Heidegger, Kant backed off "when faced with the doctrine of transcendental imagination."[11] According to Heidegger's exegesis, the "spontaneous drive" of the first draft postulated that the imagination was the "third faculty" (next to sensibility and understanding), that it founded the regimen of the One and thus ensured the possibility of ontological knowledge. Heidegger goes on to reproach Kant for renouncing to further pursue the exploration of this "unknown root" of the essence of Man and for having instead brought the imagination back to a simple operation of the understanding. As Heidegger wrote, "Kant noticed the unknown and was forced to back-up. Not only did the transcendental imagination frighten him. Along the way, he grew increasingly sensitive to the prestige of pure reason as such."[12]

In my opinion, giving up on resorting to the positivity of a third faculty (namely, the imagination) and reducing the problem of the One to that of a simple *operation* of the understanding attests, on the contrary, to Kant's critical firmness. It shows his refusal to yield to the aesthetic prestige of ontologies of Presence. The "prestige of pure reason" can very well name that firmness as he was faced with the great Temptation. What is also true is how Kant also saw the real peril as lying precisely there. He was forced to recognize the crucial meaning of the void, as much on the side of the transcendental subject as on that of the object=x. For the first time really, Kant was the one to shed light on the avenues of a subtractive ontology, far from any negative theology.

Is this to say that Kant's undertaking fully succeeded in the end? I for one would say no. Anyone can see traces subsisting in his philosophy of how the origin of the deduction resides in the theory of linkage. In fact, Kant assigns a foundational function to the *relationship* between two voids. He does so as a last resort because what he intends to found is the 'there are' of objects, the objectivity of objects. It alone manages to sustain the unfolding of a categorical connection specific to manifold representations. For Kant, the object remains the sole and unique name of the One in representation. The synthetic unity of consciousness is what I require not only for knowing an object, but it is a condition "under which every intuition must stand in order *to become an object for me*. For otherwise, in the absence of this synthesis the manifold would *not* be united in one consciousness."[13] The task of ordering the theory to the knowledge of universal relations (which is its epistemological aim) forces the power of the count-as-one to have representable objects as a result. It splits the void according to the general idea of the subject/object relationship, which remains the unshakeable framework of ontology itself.

So the Kantian Critique hesitates at the threshold of this ultimate gesture. It posits that the relation is not, and that the nonbeing of the relation is *of another kind* than the nonbeing of the One. Hence, it is impossible to lay out an identical symmetry between the void of the count-as-one (the transcendental subject) and the void as the name of Being (object=x). Naturally this gesture would also posit that the object is not the category by which the thought of a being of representations opens up. Without succumbing outright to Hume's skepticism, this thought would instead accept the dissolution of object and linkage in pure multiple presentation.

Yet Kant is a philosopher of rigor and extreme scruples.

In his desire to ground the universality of relations, he yet spotted the unthinkable abyss opening up between the withdrawn transcendental object and the absolute unity of original apperception, between the place of the being of connection and the function of the One. The hesitations and repentances with which the major differences between the two editions of the *Critique of Pure Reason* stand as witness are differences that have to do with the transcendental subject. But they do not have the concept of the imagination as a center of gravity. They are the cost of a difficult relationship between the narrowness of the premises (examining the form of judgment) and the extension of their consequences (the void as a point of Being). Clearly, and Heidegger this time devotes a decisive exegesis to it, the notion of object bears this difficulty. Kant busied himself needlessly here with a notion that should be dissolved by the operations of ontology irrespective of its pertinence for a critical doctrine of linkage.

In turn, Kant takes up the problem from another angle. It is the perspective from the abyss opened in Being in the wake of the void's double naming—according to the subject and the object. He does so by asking whence and how these two voids may be counted as one. A completely other apparatus is then required. In fact, instead of the situation epistemology affords, he needs something else. The essential challenge of the *Critique of Practical Reason* is to determine this co-belonging of the subject's void and the object's void to a single sphere of Being. Kant calls it the "supersensible." Far from being the metaphysical, 'regression' commentators have at times deciphered it as containing, the second Critique is from this angle a *necessary* dialectical resumption of the ontological impasses of the first. Its aim is to count something as one in another situation (that of willed action). In the cognitive situation, that "something" is what had remained enigmatic with respect to the two absences.

Still, Kant's powerful ontological intuitions have been captive of an order of knowledge in which the starting point is limited to forms of judgment (which is, it must be said, the lowest level of thinking activity). In the terms of localization, his notions are captive of a thought of the Subject which has made that thought a constitution protocol, whereas it is but a result at best.

Meanwhile, that the question of the Subject might be that of identity, and therefore of the One, is acceptable provided it be conceived not as the empty center of a transcendental field, but as the operatory unity of a manifold of identity effectuations—or as the *multiple ways of being identical to oneself.*

A Subject is the group of multiple possibilities of the One.

We are going to illustrate this thesis relative to a Subject's logic, instead of to its being. This will be done by means of an exercise allowing us to revisit several notions from category theory (or topos theory)—notions as primordial as they are elementary.

13

GROUP, CATEGORY, SUBJECT

First, let us recall what a group is. This algebraic structure is omnipresent in every sector of contemporary science, including the human sciences as shown in Claude Lévi-Strauss's efforts to codify kinship ties by means of combinatory groups.

This notion deals with a set endowed with an operation marked "+." It is an operation having three properties:

1. It is associative. What this means is only that were you to add element b to element c, and add element a to the result, you would obtain the same result as if you were to take the single element c and add it to the sum of a and b. This is written as:

$$a + (b + c) = (a + b) + c$$

Basically, associativity ensures the total sum of a + b + c to be indifferent to the time of operatory acts, so long as one is held to literal succession as such, that is, to the spatial order of the marks. Whether you begin by calculating b + c so as to add a *to its left* , or you begin by calculating a + b so as to add c *to its right,* you are in the process of calculating a + b + c. Associativity asserts that the result will be the same.

2. There exists a neutral element. Let an element of the basic set, say e, that is such that for *every* element a of the set, we have:

$$a + e = e + a = a$$

The existence of a neutral e, such that a + e = a, irrespective of what the element a of the group is, singularizes a pure and simple *null action* in the operatory field. This is what allows what there is to be, despite its being marked. The addition of e to some such element brings the element back as if nothing had happened. We know this from the number zero. Irrespective of the number to which it is added, zero gives that number. Zero is the neutral element of the group of relative natural numbers.

3. For every element a, there exists an element a' of the group such that, when added to a, it gives the neutral element. Thus, a' is called the "inverse of a." This time the formula is as follows:

$$a + a' = c$$

In this inverse, we can recognize, for example, the negative number –4. When added to +4, it does indeed give zero.

Generally speaking, the existence of the inverse of an element registers a classical kind of symmetry in the operatory field. Its central point is precisely the null action (the neutral element). No matter how "big" a is, or how "remote" from the neutral element e we would like it to be, there is always a' located in some sense at the other extremity in the space of marks. Its operatory effect brings element a back to the domain's ineffective center of gravity. That the neutral element e is necessarily its own inverse (e + e = e) does stabilize it as the central inertia of the symmetry domain.

The three major operatory properties defining a group structure (associativity, the neutral element, and existence of the inverse), respectively exhibit the submission of operatory acts to the literal order, the One of the null act and general symmetry. They therefore justify that this structure is in some sense the most elementary matrix of what algebra lays out as a thought.

However, the deepest essence of the group, that which legitimizes its omnipresence, is not reached by these operatory definitions. Through the latter, we remain excessively enslaved to the passive idea whereby objects or elements are what the operation operates "upon." Technically this means we assume a set of terms (the a, b, c, etc.), as if in the background of the concept, which are disjointed in their operation (the +), and in which pure manifold does not enter as such into what the concept bears of real significance.

Let us assert that a great concept in algebra (this does not necessarily obtain regarding a great concept in topology) is more encumbered than illuminated by mentioning the underlying set that is presupposed in its definition. It is really the action of operating and its major offices (detemporalization, nullity, and symmetry) which count. Its manifold marks are only a mediation.

Actually, the specifically categorical definition of a group is the only one able to reveal its essence. By contrast, the set-theoretic machinery thickens it. It makes it opaque.

This is completely natural. As we have already emphasized, categorical thinking is not ontological. It is essentially logical. It is not a proposal on universes, but an ordered description of possible universes. What is an algebraic structure if not a way to stabilize an operatory possibility with a definition? In basic algebra, every ontological option legitimately remains in abeyance. Whence we are led to emphasize that the concept's clarity requires a minimal, alleviated ontology—a slim ontology. Or what Jean Toussaint Desanti calls an "extrinsic" ontology.[1] This really means an onto-logy given from the angle of its "-*logy*," its possible form, that is, its *morpho-logy*: its morphisms. Category theory is suited to this type of donation.

In category theory, the initial data are particularly meager. We merely dispose of undifferentiated objects (in fact, simple letters deprived of any interiority) and of "arrows" (or morphisms) "going" from one object to another. Basically, the only material we have is oriented relations. A linkage (the arrow) has its source in one object and target in another.

Granted, the aim is ultimately for the "objects" to become mathematical structures and the "arrows" the connection between these structures. But the purely logical initial grasping renders the determination of an object's sense entirely extrinsic or positional. It all depends on what we can learn from the arrows going toward that object (whose object is the target), or of those coming from it (whose object is the source). An object is but the marking of a network of actions, a cluster of connections. Relation precedes Being. This is why at this point of our inquiry we have established ourselves in logic, and not ontology. It is not a determined universe of thought we are formalizing, but the formal possibility of a universe.

The stabilization of a possible universe of this type is reduced to a few elementary prescriptions:

1. Two arrows "following one other" (the target of one is the source of the other) makes up a composition. If f and g are two arrows in this situation, one notes their composition: $g \circ f$. This simply means that if object a is linked to object b (by f) and b to object c (by g), then a is linked to c (by $g \circ f$).

2. The composition of arrows is associative, which means that:

$$f \circ (g \circ h) = (f \circ g) \circ h$$

3. For every object, say a, there is an "identical" arrow $id(a)$ associated with it. It goes from a "toward" a (source and target are a), which is typically a null action. If you compose an arrow with source a and the identity arrow $id(a)$, you have "the same action" as f:

$$f \circ id(a) = f$$

It is also possible to say that the identity arrow is a neutral element in the operation of an arrow composition.

Notice how two of the operatory dispositions by which a group is defined (associativity and the neutral element) are immediately required so as to stabilize the presentation of a categorical universe. As such, these dispositions vouch for their primitive character regarding the specification of some such operatory datum, of a simple abstract possibility for thought. Associativity is only the detemporalization of gestures, that is, the establishment of the rule of the letter. On the other hand, the null action is the One's minimal power. "One" object (a letter) generates its own stopping point for actions—as its own inertia, and as a pure and ineffective identity to self.

In categorical thought, this figure of identity, albeit necessary, is not essential. It is only the identity of inertia, whereas the active "specification" of identity is evidently the goal of any logical investigation. Under what relational conditions, under what effective actions (or morphisms) can "two" objects be declared really "the same?" In categorical thought, being the same means when any identity is ultimately extrinsic. Let us use "the same" for a thought inhabiting a possible categorical universe and for the inhabitant of a category who is only cognizant of oriented relations between indeterminate objects.

The key concept, then, is isomorphism. Two objects are categorically indiscernible, albeit 'ontologically' (that is, literally) distinct, if there exists a morphism (an arrow) from one to another which is an isomorphism.

What is an isomorphism? It is a reversible arrow in the following sense. Say an arrow i operates from a toward b. There exists an arrow j operating from b toward a (in the inverse sense of i). In a composition with i, the action is null (an identical arrow). We thus have:

$$i \circ j = \mathrm{id}(a), \text{ and } j \circ i = \mathrm{Id}(b).$$

The existence of the i, j pair between a and b allows us to state that a and b are isomorphic objects and that they are (categorically or logically) identical.

The possibility of canceling out an inversion "identifies" the literally distinct objects a and b. This is not entirely self-evident. Illustrating the fact that our inhabitant of the category "sees" object a and b as "identical," implies that for her an object which has no "inside" is identified exclusively by the arrows in relation to which it is either the source or target.

It can be shown without too much trouble that if there exists an isomorphism between a and b, then the network of arrows having object a as its source and target prescribes an absolutely parallel network of arrows having b as source or target, and conversely. As such, the external—or logical—determinations of two objects, identifying them *in relational terms,* are formally the same.

But you will surely have noted in passing that the existence of an "inverse" arrow strongly recalls the existence of the inverse of an element in the definition of a group.

The characteristics of minimal operatory stability in a category are associativity and the identity arrow. Were we now to add the definition of what is the purely relational identity of two objects (inversion of an arrow, isomorphism), the group's structural determinations would be returned to us.

The group concept is thus co-present in thought once the most general investigation of what a possible universe can logically be is undertaken: a universe solely defined through anonymous objects and oriented relations between those objects.

Thereupon, it should not be astonishing to discover the categorical definition of a group is particularly simple and revealing:

A group is a category that has a single object in which every arrow is an isomorphism.

The single object, say *G,* is only the group name, its instituting letter. The arrows (all of them from *G* to *G*) are its real acting principles, its operatory substance.

Associativity is ensured by the fact that it is categorical for the composition of arrows. In the algebraic set-theoretic version, what was written a + b (an operation between elements) is now written g o f, where *f* and *g* are arrows from G toward G. In other words, what used to be written as:

$$a + (b + c) = (a + b) + c$$

is now written as:

$$f \ o \ (g \ o \ h) = (f \ o \ g) \ o \ h.$$

The neutral element is the identical arrow of the sole object *G,* that is, *id*(G). Indeed, for every arrow of category *G,* we have: f o id(G) = *f.*

Finally, as all of the arrows are isomorphisms, each possesses an inverse arrow, say *f'* such that f o f' = *id*(G). Since id(G) plays the role of neutral element, this covers the third structural property of a group operation.

In every category *G* having a single object of which all of the "elements" are arrows, we can recognize a group with the following properties:

- The "elements" are arrows;
- The operation is an arrow composition;
- The neutral element is the identity arrow of G, and;
- The inverse of an element is the inverse arrow determining the first arrow as an isomorphism.

Yet recognizing this group, which alone is a kind of wild isomorphism between the algebraic concept of group and a certain kind of possible categorical universe, is not what is most interesting.

What is interesting is the following: the categorical definition of group *G* makes *G* appear as *the set of the different ways in which object-letter G is identical to itself.*

We have seen how identity in categorical thought is fundamentally isomorphism. Since all of the arrows of a group *G* are isomorphisms, each of these arrows registers an identity of *G* to itself. Accordingly, two different arrows represent the difference between two self-identities.

This means that the real purpose of a group is to set the plurality of identity. Now we hold the principle for thought of this concept's ubiquity.

Among the different ways of being identical to oneself, there is "inert" identity, that is, the null action *id*(G). What a group indicates is that this identity is but the degree zero of identity, its immobile figure and at rest. The other arrows of the group are dynamic identities. They are the active ways of being identical to oneself. They show that what vouches for the One of entity *G*, over and above its empty literal fixity, is the plurality of its internal connections through which it produces isomorphism and, thus, the multiple ways of manifesting its own identity.

The more you have different arrows, the more the letter *G* proves to be the name of a dense entity, a complex network of differentiated identities.

At this point we ought to evoke Plato's *Sophist,* and the five supreme genera of all intelligibility. Being, pure Being and purely empty Being, is letter *G*, which is but the literal index of the One-that-is. Rest is *id*(G), inert identity to oneself as the stopping-point of all action. Movement is the arrows, noninert isomorphisms plaiting *G* in the active manifestation of its identity. Last in Plato's list is the dialectic of the Same and the Other, which is made explicit in the difference of arrows. For this difference, in terms of difference, vouches for the Other. But this Other is also only the differentiating work of the Same because every "other," being an isomorphism, is nonetheless a figure of the Same.

Let us say that a group is the minimal presentation of the intelligible in terms of the otherness of the same.

What does the coupling of arrows and the idea that every arrow might be associated with its inverse mean for thought? It means that given a singular way of being identical to oneself an arrow prescribes its "being in the mirror," or its symmetry, in the same stroke. Since the dawn of philosophy, it is well known that identity is not so much linked to a mimetic disposition as to a specular one. Every identity imposes its reversal as the other identity that doubles as well as annuls it. The primordial datum is not so much a self-identity distinct from any other, than the couple of two symmetrical identifications, the upshot of which is that when taking them together we fall back upon the identity of inertia. This is expressed as a matheme in the equation f o f' = $id(G)$.

Finally, the concept of group, rendered in its essence by the categorical definition (an object and its isomorphisms), is a real thought on identity:

a) The inaugural One is never but the empty point of a letter, *G*, deprived of any interiority.

b) This letter is affected by an inactive, inert identity, a pure stopping point for any immanent action *id(G)*. It is a relation to itself which lets empty literality be.

c) There is a plurality of identity-to-self modes. Every mode is fixed by an action of self toward self. It is an isomorphism that actively "informs" on identity, an arrow *f* of *G* toward *G* which performs identity in a singular fashion. Even the Same is always caught in the meshwork of the Other: this way of being identical is different from one or many others.

d) Every identity mode is coupled to another singular mode, which is its reversal, symmetry and image. The composition of the couple returns an identity of inertia. Coupled to its image, an identity is just as soon ineffective.

It is now high time to say that the group works as a matheme for a thought on the Subject. It is formally adequate to what Freud and, later, Lacan attempted to record as its fleeing identity.

In the beginning, there is but a letter if we maintain that a proper name is the position of a letter-signifier in the Subject economy. As such the proper name is empty; it says nothing. The Subject presents itself instead as a plait consisting of the active figures of its identity. It is the signifying articulations wherein desire is presented, that is, the differentiated pieces of information of the initial letter wherein the same-subject ek-sists from the plurality-other of its identifications. The psychoanalytic cure is akin to unfolding the strands of the plait. It is the possibility of considering not the identity of inertia—which is only the null index of the proper name—, but of otherness itself, the plural and intertwined arrowing of immanent isomorphisms. Finally, what Freud thought, under Empedocles' sign, as the antinomy of the life and death drives, strings each tress, as it were, to the register of the double and the image. There is no linkage of desires which is not haunted specularly by a mortifying image. This coupling of the Real and the Imaginary brings what Lacan names its *caput mortuum* into being as a key ingredient of the symbolic sphere. The death's head is the emblem of the identity of inertia, akin to a tomb at which the only thing engraved is the name. Indeed, the group says it all: at death, like at its birth, there is only the vacuity of the letter left ("let a group *G*").

The cure is how we illuminate what still remains the mysterious relationship between psychoanalytical theory and its situation. Let us say that there is a psychoanalytic theory like there is a group theory. In the final analysis, psychoanalytic theory thinks the great categories unpinning the subjective tress: the letter, the identity of inertia, the meshwork of the Same by the Other, the operatory plurality of the identical, the image. Whereas the cure intends to identify a singular group, a subject-group intransitive in its existence to theory. The "unfolding" of the latter is done strand by strand.

The obstacle is that everything seems to lead one to surmise the following: a subject-group is infinite. Perhaps is it even proper to a subject that the different ways it has of being identical to itself is infinite in number. The upshot of this is that psychoanalysis is incompletable as such. The only thing we shall have of the group is what makes it suitable for the subject to bear its proper name. What the modest and radical aim of psychoanalysis is all about involves provisionally separating enough isomorphisms from the subjective tress so that the living effectivity of self-identity does not take place at the risk of a permanent return of inertial identity. Psychoanalysis lifts the infinite life of the subject-group to the height of the letter naming it. It is a vivification of the letter.

Its logical principle remains throughout that there is only a dead way of being identical to oneself, or of being "oneself," as it also contends that, in the isomorphism group of the subjective structure, there is a living infinity of such modes of the identical.

This is our luck. Yet were we here to shift the Subject and its infinite capacity to modulate and modify its identity to self over to logic's side, rather than to ontology's, it would still have us wind up with the question of determining the "scrap" of Being from which this logic is woven.

Let's put this more bluntly. How is the thesis of ontology as the mathematics of the pure manifold and that of logic as the (equally mathematical or mathematized) science of possible universes of the relation articulated? What is it that, as inconsistent as their being is, leads every world and every situation to be implacably *linked*?

This is the question on which this short book opened. In fact, the book is really but the introduction to the real difficulty. We are going to deal with it in the second volume of *L'Etre et l'événement*, in which the logical match up between the Subject and truths ends up finding its logic, as well as its ethics.

I shall only lift a corner of the veil here by introducing, as closely as possible to what has just been said and by recapitulating its steps, the concept in which everything is played out. This is the concept of appearing (*l'apparaître*).

14

BEING AND APPEARING

Let us to return to a remark, a distressingly banal one: logic is a mathematical discipline today. In less than a century, it has reached a dense complexity yielding nothing to any other living region of the mathematical sciences. There exist logical theorems, specifically in model theory. Their very arduous proofs synthesize methods originally stemming from apparently remote domains of the discipline, that is, from topology or transcendental algebra. Their power and novelty are astonishing.

What is most astonishing for (continental) philosophy, however, is the scant astonishment this state of things has aroused. For Hegel it was still quite natural to give the title of *The Science of Logic* to what is evidently a broad philosophical treatise. Its first category is Being, Being *qua* Being. In addition it contains a lengthy development earmarked to ascertain that, insofar as the concept of the infinite is concerned, mathematics is but the immediate step of its presentation and must be relieved by the movement of dialectical speculation. For Hegel, this dialectic alone is what is fully deserving of the name "logic."

The fact that mathematization finally won as far as what the identity of logic is all about has remained a challenge thrown most specifically at philosophy, which historically speaking actually set up the concept of logic and laid out its forms.

The question then becomes the following: What does logic stand for? And what does mathematics stand for regarding a situation whose destiny has seen logic be inscribed into mathematics? This inscription has determined a kind of torsion, which has itself become a question within our question. For if ever there was a discipline that demanded the conduct of its discourse be strictly logical, mathematics is certainly the one. The upshot is that logic appears to be one of the a priori conditions of mathematics. How is it possible, then, for this condition to end up as if injected into what it conditions, to the point of only being a regional disposition in it?

Surely we have to hold that the mediation between logic as a philosophical prescription and logic as a mathematical discipline is found in what it has become customary to call logic's *formal* character. It is well known that in the preface to the second edition of the *Critique of Pure Reason,* Kant attributed the path logic had taken from "the earliest times" with the character of a science. Kant also argued that since Aristotle logic had been unable to retrace or advance a single step due to the way logic "gives an exhaustive exposition and a strict proof of the formal rules of all thought."[1] Its success is entirely linked to its abstracting from all objects and consequently ignoring the great divide between what is transcendental and what is empirical.

So, we can rephrase our question in the following way, which I believe is the commonest conviction heard today: from the fact that logic is formal and not tied to any figure of the empirical donation of objects, the outcome is that its destiny is mathematical. It is such in that mathematics itself is precisely a formal theoretical activity. It is so in the sense according to which Carnap for example distinguished the formal sciences, namely, between logic and mathematics on the one hand, and the empirical sciences of which physics is the paradigm.

Nevertheless, we can notice right off the bat how this solution cannot be Kant's. Kant was constantly faithful to ontological intuitions whose destiny has already been discussed. For Kant, mathematics requires the form of temporal intuition for the genesis of arithmetic objects, and the form of spatial intuition for geometric ones. Mathematics cannot be held to be a formal discipline. This is why if logical judgments remain analytic, mathematical judgments, including the simplest ones, are synthetic. It can also be noticed that the attribute of logic's immutability since its Aristotelian foundation, which Kant winds to its formal character, covers up a double error, which has to do with history and forecasting. Kant's assessment is an historical error because he does not bear in mind the complexity of logic's history. Long ago, the Greeks barred any presupposing over logic's unity and fixity in the way Kant does.

Even more to the point, Kant entirely erases the fundamental difference in orientation between Aristotle's predicate logic and the Stoic's
propositional logic, a difference from which Claude Imbert recently
still managed to draw important consequences.[2] It also displays an
error in forecasting. In other words, despite the fact of being ensured of its mathematization, logic has not folded. It is precisely
one of the most magnificent efforts in thought to have marked the
twentieth century, during which it made only giant strides forward.

For all that, it is quite remarkable that Kant's thesis is exactly the
same as Heidegger's. Kant's was earmarked to bring out the merits of
logic as well as its limitations for the very general forms of nondescript
thought. Heidegger's approach was earmarked to something altogether
different, namely, checking off the forgetting of Being. Its formal au-
tonomy from logic is one of its main effects. In Heidegger's mind, it
is well known that logic was produced from a split between *physis* and
logos. It is the potentially nihilist sovereignty of a *logos* regarding
which Being has withdrawn. But to broach this historical determination
of logic, what does Heidegger tell us about its self-evident characteristics? Quite simply that logic is "the theory of the forms of what is
thought." Like Kant, he infers that "since the beginning logic teaches
the same thing." Formalism and immutability both seem tied and suitable to a vision of logic confining it either to what is held on one side
of the divide between the empirical and transcendental, which was
Kant's case. Or, like with Heidegger, it is the technical purpose of a
nihilistic arresting of Being in its totality.

It is difficult to ensure that the mathematization of logic is really
the consequence of its formal character. Either this thesis comes up
against the fact that mathematization has instilled logic with an im-
petuous drive, contradicting the immutability its formal character
seemingly imposes. Or this thesis assumes mathematics itself is purely
formal, which is something that would require us to ponder what it
is then that distinguishes it from logic. Admittedly, in the twentieth
century, the "logicist" project, which strove to reduce mathematics to
logic, foundered upon paradoxes and impasses as early as Frege's
groundbreaking work. So much so that, and despite being entirely
mathematized, what logic itself seems to prescribe is how the whole
body of mathematics cannot be drawn over to its terms.

We are thus led back to our question as a question. What
does it mean for thought that logic might be identifiable today as

mathematical logic? This established syntagma ought to astonish us. We should ask: What is logic after all, and what is mathematics after all, for it to be possible and even necessary to speak of mathematical logic? I reckon the reader has grasped that the conviction at work from the first strokes of this treatise is that it is impossible to start drafting an answer to this question without using a third term. Although present since the very outset, the syntagma "mathematical logic" has actually marshaled this term's absence. The third term is "ontology," that is, the science of Being *qua* Being.

For both Kant and Heidegger, Aristotle was the founder of what is understood by the word "logic." It is from this third term that Aristotle allows himself to question the formal necessity of the first principles of any discourse professing consistency. Thinking of Being, Being *qua* Being, summons us to determine what the axioms of thought are in general. This is Aristotle's central thesis in Book Gamma of the *Metaphysics*. He states: "It is upon those who seek to make a theory of first essence that equally rests the task of examining its axioms."[3] This is why the initial announcement according to which there exists a science of Being *qua* Being ends up being crossed out, instead of carried out, by legitimating the principle of noncontradiction at considerable length ("It is impossible for the same to simultaneously belong and not belong to the same and according to the same.")[4] This is followed up by the principle of the excluded middle ("It is necessary either to affirm or negate for one subject one predicate no matter what it might be.")[5] That these principles stand nowadays as logical laws can no longer be doubted. This is true even to the point wherein admission or rejection of the second principle, of the excluded middle, discriminates the two fundamental orientations of contemporary logic, the classical and intuitionist orientations. The classical orientation ends up validating reasoning by the absurd (the reductio ad absurdum), whereas the intuitionist orientation admits only constructive proofs. This assures us that Aristotle lays out logic as a binding mediation of ontology. Whoever declares the existence of a science of Being *qua* Being will be expected to explain the formal axioms of any transmissible discourse. Let us be agreed, then, that for Aristotle ontology prescribes logic.

Why? So as to understand that after acknowledging the existence of ontology, we have to take on Aristotle's second statement.

This one focuses on the difficulty, in his view, of the science of beings *qua* beings. It implies that the existent is said in several ways. But it also means that the existent is said as *pros hen,* in direction of the One, toward the One or in possible seizure of the One.[6] The thesis is that ontology cannot be constituted according to an immediate univocal hold of an object that would be presupposed to it. Beings, or the existent as such, are not exposed to thought in the form of the One, but in the equivocation of sense. Ontology should not be conceived as a science of a given or experimented object in evidence of its unity, but as a construction of unity regarding which only the direction, *pros hen,* toward the One, is given. This direction in turn is all the more uncertain in that its starting point is an irreducible equivocation. Holding the direction engages one in the construction of a unity in the design of the science of Being. The upshot of holding the direction is to presuppose that we determine the minimal conditions of the univocity not of an object, but of discourse as such. What universal and univocal principles does a consistent discourse rest upon? Agreement on this point is necessary if only for taking the direction of the One when attempting to reduce Being's initial equivocity. Logic is set exactly in the interval between Being's equivocity and the constructible univocity in regard to which this equivocity is a sign. Its formal character has to be reduced to this. To put it metaphorically, logic stands in the void that, for thought, separates the equivocal from the univocal insofar as it is a question of beings *qua* beings. This void is indexed by Aristotle to the *pros* preposition. For ontological discourse, it indicates the direction according to which discourse may constructively cross the void between the equivocal and univocal.

In the end, in as much as ontology admits equivocal sense as a starting point, it prescribes logic as an explicit exhibition of the formal laws of consistent discourse, or as a form of scrutiny over the axioms of nondescript thought.

Now observe how the choice of the equivocal as an immediate determination of beings caught in their Being, as it were, excludes with Aristotle any ontological pretension mathematics might have. For mathematics has two characteristics, both of which were perfectly recognized by Aristotle, in particular in Books B and M of the *Metaphysics.* On the one hand, it is dedicated to the univocal, which also means for Aristotle that mathematical things, the *mathematika,* are eternal, incorruptible and immobile. But this univocity is paid for, as it were, by the fact that the being of mathematical things is only, as seen in the foregoing, a pseudo-being, indeed a fiction. Mathematics cannot open any access whatsoever to the determination of beings *qua* beings. As bound to pure logic, mathematics is a *fictioning* of eternity. Its destiny is like fiction's own, that is, it is not ontological but aesthetic. From the fact that ontology is rooted in equivocity, the upshot is that logic is prescribed as the formal science of the principles of consistent discourse and that mathematical univocity is but a rigorous form of aesthetics. Such is the Aristotelian knot of ontology, logic and mathematics.

There are several ways to undo the knot. To a certain extent, all of them are Platonic. All of them, by postulating that Being should be sayable in a single sense, establish mathematical univocity again as ontology's paradigm—at least provisionally. Especially, all of them restore the pertinence for mathematics of the category of truth, which necessarily mediates between the act of thought and the act of Being. This restoration of the theme of mathematical truth contrasts with Aristotle's relativist and aesthetic conviction. For Aristotle, the de-ontologization of mathematics draws the beautiful into the place of the true.

We could also put it otherwise. Whosoever thinks mathematics pertains to an order of rigorous fiction, for example, linguistic fiction, ends up changing it into an aesthetics of pure thought. The Aristotelian is involved in such a thought. Whosoever thinks mathematics touches upon Being is a Platonist. This is why the opposition between Plato and Aristotle has been one of the major motifs of this book.

Notice also how logic cannot hold the same place in both options. What lends logic its force in an Aristotelian's point of view, including from the standpoint of mathematics? Logic is purely formal and absolutely universal. It presupposes nothing by way of ontological determinations and it is linked to the consistency of discourse in general. These reasons underscore why logic is the obligatory norm in the passage from Being's equivocity to the unity toward which its equivocity is the sign. For a Platonist these characteristics are weaknesses. In her eyes, mathematics conceives of the kind of idealities whose status with respect to Being is undeniable. By contrast, logic remains empty. For logic to be raised, it has to be mathematized enough to share with mathematics the ontological dignity the Platonist recognizes in the *mathematika*. Whereas for the Aristotelian, logic's purely formal dimension is precisely what keeps it from being caught in the aesthetic mirage of the *mathematika,* that is, inexistent quasi-objects. Logic's principled, language-based and nonobjective character is what founds its discursive interest for ontology itself.

Say that the Platonic knot is an ontological promotion of mathematics which deposes logic, whereas the Aristotelian knot is an ontological prescription of logic that deposes mathematics.

It could then be said that the position I ready myself to uphold is, to speak like Robespierre stigmatizing the factions, both *ultra-*Platonic and *citra*-Platonic.[7]

It is *ultra*-Platonic in as much as we assert once again that ontology is nothing but mathematics itself, and thus push recognition of the ontological dignity of mathematics to its extreme. Whatever can be said rationally of Being *qua* Being, of Being as deprived of all qualities or predicates other than the simple fact of being exposed to thought as an existent, is said—or rather written—as pure mathematics. The effective history of ontology coincides exactly with the history of mathematics.

On the other hand, our position is *citra*-Platonic in as much as we shall not have to take on the destitution of logic. Actually, we go on to see that by presupposing the radical identity between ontology and mathematics, logic can be identified differently to a formal discipline regulating the usage of consistent discourse. Logic can be torn from its grammatical status. It can be separated from what is called the "linguistic turn of contemporary philosophy."

It must be said that this turn is essentially anti-Platonic. For the Socrates of the *Cratylus,* the maxim is that we philosophers leave off from things and not words. Moreover, what can be said is that we leave off from mathematics and not formal logic. Let no one who is not a geometer enter here. All in all, the linguistic turn only ensures the tyranny of Anglo-Saxon ordinary language philosophy. Therefore, any act of reconsidering the linguistic turn amounts to assuming that mathematical thought, or mathematics as thought, deals with the Real—and not words.

For a long time I had believed this superceding of Platonism involved a destitution of formal logic as the royal path by which we have access to rational languages. According, and so deeply French in this respect, I rallied to the suspicion that, in the minds of Poincaré and Brunschvicg, was cast upon what they called "logistics." It was only at the price of long, arid work on the most recent formulations of logic and by grasping their mathematical correlations that I have come to an understanding on the matter. This is work that I have barely completed and of which I have only given a profile or program here. But if we manage to shed some light on how mathematics is the science of Being *qua* Being, and if we lay it out not as a syntactic norm, but as the immanent characteristic of possible universes, logic will finally be shifted to standing beneath an ontological prescription—and not a linguistic one. Doubtless this prescription takes up again with the Aristotelian gesture, although now its purpose is completely changed.

We can then do full justice—justice dispensed, if I dare say, by Being itself—to the aforementioned enigmatic syntagma of "mathematical logic." In unfolded form, it should now be called the plurality of *logics* instituted by an ontological decision.

That ontology might be historically accomplished as mathematics is the initial motif of my book, *L'Être et l'événement*. I have neither the intention nor the possibility of reiterating its body of arguments here, though I did have a chance to recall its central maxims at the beginning of these briefings.

What is important for us insofar as logic is concerned is a derived thesis. It is a theorem that can be deductively inferred from the fundamental axioms of Set Theory, and, therefore, from the principles of the ontology of the manifold. Its most common rendition is: "There does not exist a set of all sets." This nonexistence means that thought is not able to uphold the assumption by which a multiple, and hence an existent, might be the gathering together of all thinkable existents without at the same provoking its collapse. In reference to the category of totality, this fundamental theorem designates the inexistence of Being as a whole. In some respects, and according to a transposition of physics to metaphysics, it settles the first antinomy of pure reason in favor of its antithesis: "The world has no beginning, and no limits in space; it is infinite as regards both time and space."[8] Naturally, it is a matter here neither of time nor space, nor even of the infinite. As said and repeated, the latter is but an actual determination, simple and unproblematic, of beings in general. Let us put it instead as: it is impossible for thought to apprehend as multiple something that would consist of all beings. Thought fails at the very point Heidegger calls the "existent in totality" (*l'étant en totalité*). The fact that this statement might be a theorem once we accept that ontology *is* mathematics, and that the properties of the existent *qua* existent are demonstrable, means that it ought to be understood in its strongest sense: *it is an essential property of the existent qua existent that there cannot exist an existent as a whole, so long as existents are thought from their being-ness alone (étantité).*

A crucial consequence of this property is that any ontological investigation is irremediably *local*. There cannot exist a proof or intuition dealing with Being *qua* the totality of existents, or even Being *qua* a general place wherein existents are arranged. This powerlessness does not only consist of de facto inaccessibility, or a limit transcending reason's capabilities. Quite the opposite, it is reason itself that determines the impossibility of the whole as an intrinsic property of the manifold-being-ness (*étantité*) of the existent.

Let us put it simply. The determination in thought of what can be rationally said of the existent *qua* existent, and thus of the pure manifold, always assumes the place of this determination to be not the whole of Being, but a particular existent. This assumption holds even though it is on the scale of infinitely many infinites.

Being is only exposed to thought as a local site of its untotalizable unfolding.

But this localization of the site of ontological thought, which I call a "situation" in *L'Être et l'événement*, affects Being as soon as, in terms of a pure manifold, it no longer contains anything in its being by which to explain the limits of the site wherein it is exposed. The existent *qua* existent is a multiple, a pure manifold, a manifold without-One or a manifold of manifolds. It shares this determination with *all the other existents*. But "all the other existents" does not exist and has no being. The upshot of this is that, in as much as this determination is given, it is only given in a site or situation that when thought in its being *qua* being is a manifold-existent. This is not a situation of the existent's ontological generality, which would be the inexistent whole of existents sharing the determination of their being as pure manifold. The existent can only assert its "beingness" (*étantité*) at one site whose local character cannot be inferred from this existent as such.

What links a being to the constraint of a local or situated exposure of its manifold-being is something we call this existent's *"appearing."* It is the existent's being to appear insofar as Being as a whole does not exist. Every being is being-there. This is the essence of appearing. Appearing is the site, the "there" (*là*) of the multiple-existent insofar as it is thought in its being. Appearing in no way depends on space or time, or more generally on a transcendental field. It does not depend on a Subject whose constitution would be presupposed. The manifold-being does not appear *for* a Subject. Instead, it is more in line with the essence of the existent to appear. As soon as it falls short of being localized according to the whole, it has to assert its manifold-being from the point of view of a non-whole, that is, *of another particular existent* determining the being of the there of being-there.

Appearing is an intrinsic determination of Being. The localization of the existent, which is its appearing, involves another particular being: its site or situation. This is why it can be seen immediately that appearing is as such what connects or reconnects an existent to its site. The essence of appearing is the relation.

Now, the existent *qua* existent is absolutely unbound. This is a fundamental characteristic of the pure manifold as it is thought in Set Theory. There are only multiplicities and nothing else. None of them on their own is connected to another. In Set Theory even functions have to be thought as pure multiplicities or manifolds. This is why we identify them by their graph. The "beingness" of the existent does not presuppose anything else than its immanent composition, that is, that it might be a manifold of manifolds. Strictly speaking, this excludes the possibility that there might be a being of the relation. When thought as such, and therefore purely generically, Being is subtracted from any connection.

It belongs to Being to appear, and thus to be a singular existent. In as much as it does, Being can only do so by undergoing a primordial connection with the existent situating it. Appearing— and not Being as such—is what superimposes the world of the relation to ontological disconnection.

This is how light is shed upon a kind of empirical self-evidence. By combining ultra-Platonism and citra-Platonism, we end up with a reversal of Platonism. Platonism seems to state that appearance is equivocal, mobile, evasive, and unthinkable. It seems to say that ideality, including mathematics, is what is stable, univocal and exposed to thought. But as moderns we can argue in favor of some opposing evidence. The immediate world, the world of appearances, is the one to be always given as solid, connected and consistent. It is a world of the relation and cohesion, in which our bearings and customs lie. It is a world in which Being is, all in all, captive of Being-there. Being in itself, thought as the mathematicity of pure manifold—or even as the physics of quanta—is what proves most anarchic, neutral, inconsistent, and indifferent to what signifies. Being itself is what maintains no relationship with anything that is not it.

To be sure, Kant had already started off by setting the phenomenal world as always connected and consistent. The question this world poses for us already consisted for him of reversing Plato. It is not the inconsistency of representation that creates a problem, but its cohesion. What must be explained is that appearing constitutes an always linked and connected world. There is no contesting that the *Critique of Pure Reason* interrogated the logic of appearing.

Having said this, Kant inferred from the condition of this logic of appearing that Being itself remains unknowable for us. Consequently, he set down the impossibility of obtaining a rational ontology. For Kant—and this is a nodal point that is neither Aristotelian nor Platonic—the logic of appearing is what deposes ontology.

For us, conversely, ontology exists as a science. Being itself befalls to the transparency of the thinkable in mathematics. But there is a catch, which is that this transparency grants Being merely with the pure multiple, or manifold, and its sense-deprived rationality. Being *qua* Being is caught in the infinite task of its knowledge, which is the historicity of mathematics. Accordingly we can say that appearing as such is what ends up imposing the possibility that there might be a single logic, since it is appearing that fixes the "there" of Being-there as a relation. The ontological plinth is but the tendentious inconsistency of the pure multiple, the manifold, as it is thought in mathematics.

This should begin to shed light on our initial problem. Let logic be what makes the science of appearing *an intrinsic dimension of Being*. On the other hand, let mathematics be the science of

Being *qua* Being. In as much as appearing, that is, the relation, is a constraint that affects Being, it must be the case that the science of appearing itself be a component of the science of Being, and hence of mathematics. Logic is required to be mathematical logic. But so long as mathematics apprehends Being according to its being—on this side of appearing and of its fundamental disconnection—mathematics is also required to steer clear of logic and avoid getting mixed up with it.

Let us posit that logic is from within mathematics the movement of thought explaining the being of appearing, that is, of what affects Being as long as it is Being-there.

Appearing is nothing but the logic of a situation, which is always in its being *this* situation. And logic as science restitutes the logic of appearing as a theory of situational cohesion in general. This is why it is not the formal science of discourse, but the science of possible universes as thought according to the cohesion of appearing, which is itself the intrinsic determination of the existent *qua* existent's disconnection.

By now we have moved very close to Leibniz. Logic is what applies for every possible universe. It is a coherence principle required for every existent once it appears. But we are also quite far from Leibniz. For what is thought in its being is not regulated by any harmony or principle of reason. On the contrary, it is disseminated in the inconsistent and reasonless manifold.

This is the point at which we can ask where and how—from within mathematics—is the mathematical status of logic made clearer. The answer is: the mathematical theory of possible universes, the general theory of the cohesion of Being-there, or even more the theory of the relational consistency of appearing.

In this respect, we cannot remain satisfied with logic's formalization in the way it was carried out from Boole and Frege down through Gödel, Tarski, and Kleene's sophisticated developments. Although admirable, this formalization is nonetheless a simple aftereffect of Aristotle's initial constructions, to which the predicate calculus and proof theory correspond. It also follows from the Stoics's constructions, to which the propositional calculus and modal logic correspond. This logical formalism assumes, logic to be a construction of formal languages as the Greeks themselves did. It consolidates the idea that logic is but the hard core of a generalized rational

grammar. It is inscribed in philosophy's linguistic turn. Its formalization reckons it can make do without the ontological prescription.
It lacks the organic identity of logic, appearing, or Being-there. Its
mathematized appearance is derived and external, for it is but a
computable literalization and accidental univocity. In a nutshell, in
this figure of logic, mathematization is but formalization. Now, in
no way does mathematics have formalization as its essence. Mathematics is a thought, a thought of Being *qua* Being. Its formal
transparency is the direct outcome of the thesis according to which
Being is absolutely univocal. Mathematical writing is the transcription and inscription of this univocity.

If logic might be said to be mathematical in the full sense of
the term, two conditions would have to be met which the simple
theory of formal languages is very far from reuniting.

First condition: Logic has to emerge from within mathematical
movement itself, and not as a will to impose a linguistic framework
from without onto mathematical activity. Not even the ontological
theory of sets was born with Cantor from a general and external
purpose, but from problems internal to topology and the classifica-
tion of real numbers. As for its mathematicity, logic can only be
made clearer if the gesture instituting and distinguishing it ends up
reproducing the fundamental motif occupying us in one effective
movement: that appearing be an intrinsic dimension of Being, and
logic, which is the science of appearing, be called and summoned
from within the science of Being and, thus, of mathematics.

Second condition: Logic must not be bolted to grammatical and
linguistic analyses. Its first question must not be that of propositions, judgments, or predicates. Logic should first and foremost be
a mathematical thought of what a universe of relations is; or of
what a possible situation of Being is, insofar as it is thought in its
relational coherence; or of what Being-there is, as the connected
essence of Being's ineluctable localization.

It turns out then that a contemporary theory of logic, the speci-
ficity of which we have already more than caught a glimpse of,
abides by these two conditions. It breaks with the linguistic, formal-
ist, and axiomatic protocols within which modern logic as a whole
had appeared to be confined. This theory, let us name it again one
last time, is the "theory of categories." Its achievement is the theory

of *toposes*. The latter is named well, for what is at stake here is Being's place.

Category theory was drafted by Eilenberg and Mac Lane in the 1940s stemming from needs that were immanent to modern geometric algebra. This satisfies the first condition. Beneath the concept of topos, category theory unfolds a thought of what is an acceptable or possible universe for some such mathematical situation to be localized in it. The logical dimension of this presentation of universes is entirely immanent to it. It is given as a mathematically assignable characteristic of the universe—and not as a linguistic and formal exteriority.

It is surely not a matter here, not anymore in fact than it was in the preceding chapters, of entering into the technique of what is commonly called the "categorical presentation of logic," or "elementary *topos* theory." However, I do intend to highlight three features that are adequate to covering the philosophical questions we have tackled.

1. The theory *of toposes* is descriptive and not really axiomatic. The classical axioms of Set Theory lay out an untotalizable universe of the thought of pure manifold. Say that Set Theory is an ontological decision. *Topos* theory defines the conditions beneath which it is acceptable to speak of a universe for thought, based on the absolutely impoverished concept of relation-in-general. Consequently, we may also speak of the localization of a situation of Being. To spin a Leibnizian metaphor: Set Theory creates "through fulgurations," a singular universe in which what there is thought according to its pure "there is." *Topos* theory describes the possible universes and their rules of possibility. It is like inspecting the possible universes Leibniz conceived as contained in God's Understanding, as it were. This is why it is not a mathematics of Being but a mathematical logic.

2. The purely logical operators are not presented in a *topos* as linguistic forms. They are constituents of the universe that are in no way formally distinguished from the other constituents. I have said that a category, hence a *topos,* is defined from a completely general and elementary notion:

either an oriented relation from object *a* toward object *b,* called an arrow or morphism. In a *topos,* negation, conjunction, disjunction, implication, and universal and existential quantifiers, are but arrows, whose definitions are given. Truth itself is but an arrow of the *topos,* the truth-morphism. As such, logic is nothing else than a particular power of localization immanent to such or such a possible universe.

3. *Topos* theory explains the plurality of possible logics. This point is crucial. Indeed, if Being's local appearing is intransitive to its being, there is no reason why logic, which is the thought of appearing, should be unique. The linkage form of appearing, which is the manifestation of the "there" of Being-there, is itself a manifold. *Topos* theory allows us to understand in depth from the mathematicity of possible universes where and how logical variability, which is the contingent variability of appearing as well, is marked in relation to the strict and necessary univocity of Being-multiple. For example, there can be classical *toposes* that validate internally the excluded middle or the equivalence of the double negation with affirmation. There can be non-classical logics as well, which do not validate either of these two principles.

For these reasons and several others, on which only a mathematical follow-up of the construction of the topos concept will shed some clarifying light, we can assert this theory to be indeed mathematical logic as such. Within ontology, it is the science of appearing, that is, the science of what signifies that every truth of Being is irremediably a local truth.

Besides, *topos* theory culminates in magnificent theorems on the local and the global. It elaborates a kind of truth geometry. It gives an integrally rational sense to the concept of local truth. What we can read in it, as in an open theorem, as it were, is that the science of appearing is also and at the same time the science of Being *qua* Being, in the specific inflexion inflicted upon it by the place earmarking a truth for it.

Accordingly, Aristotle's wish for an ontologically prescribed logic has been achieved—although not from the equivocity of Being, but its univocity. What draws philosophy—under the condition of mathematics—to rethink Being according to what is, in my view, a contemporary program, is the task of understanding how it is possible for a situation of nondescript Being to be both a pure manifold at the selvage of inconsistency, and an intrinsic and solid linkage of its appearing.

Only then shall we know why, when a novelty is shown, when the Being beneath our eyes seems to shift its configuration, that this always occurs for want of appearing—in a local collapse of Being's consistency and thus in a provisional termination of any logic. For what then surfaces, what displaces or revokes logic from the place, is Being itself in its redoubtable and creative inconsistency. It is Being in its void, which is the non-place of every place.

This is what I call an "event." All in all, it lies for thought at the inner juncture of mathematics and mathematical logic. The event occurs when the logic of appearing is no longer apt to localize the manifold-being of which it is in possession. As Mallarmé would say, at that point one is then in the waters of the wave in which reality as a whole dissolves. Yet one also finds oneself where there is a chance for something to emerge, as far away as where a place might fuse with the beyond, that is, in the advent of another logical place, one both bright and cold, a Constellation.

ANNEX

Published essays used as material
from which this book was drafted

A. Badiou, "L'ontologie implicite de Spinoza." In *Spinoza: puissance et ontologie.* Ed. Myriam Revault d'Allones. Paris: Kimé, 1994.

A. Badiou, "Platon et/ou Aristote-Leibniz. Théorie des ensembles et théories des *topoi* sous l'oeil du philosophe." In *L'Objectivité mathématique.* Ed. Jean-Michel Salanskis. Paris: Masson, 1995.

A. Badiou, "De la vie comme nom de l'être." In *Deleuze.* Ed. Françoise Proust. Paris: PUF (Collège international de philosophie), 1998.

A. Badiou, "Logique et ontologie." *Bulletin de la Société française de philosophie* (1998).

NOTES

Introduction: Alain Badiou: Back to the Mathematical Line

1. Alain Badiou, "La Subversion infinitésimale," in *Cahiers pour l'analyse* (Paris: École normale supérieure, 1968), 9: 118–137; "Marque et manque : à propos du zéro," in *Cahiers pour l'analyse* (Paris: École normale supérieure, 1969) 10: 150–173; and *Le Concept de modèle. Introduction à une épistémologie matérialiste des mathématiques* (Paris: Maspero, 1972).

2. Perry Anderson, "Dégringolade," *London Review of Books* 16, no. 17 (September 2, 2004).

3. All of Badiou's writings on Samuel Beckett have now been collected in English in Alain Badiou, *On Beckett,* trans. A. Toscano, (London: Clinamen Press, 2004).

4. A. Badiou, *Ethics: An Essay on the Understanding of Evil,* trans. Peter Hallward, (New York: Verso, 2001).

5. Jean-Toussaint Dessanti, "A Propos de l'ontologie intrinsèque d'Alain Badiou," in *Les Temps modernes* 526 (mai 1990) : 61–71. As backup for his doctoral seminar at Jussieu/Université Paris VI, nonetheless registered under Université Paris VIII, Badiou drafted a 153 page manuscript: *Topos ou Logiques de l'onto-logie. Une introduction pour philosophes.* tome 1 (unpublished, 1991).

6. Gilles Deleuze and Felix Guattari were blunter in the criticism they addressed to Badiou's reading of Set Theory when underscoring how "even mathematicians have had enough with Set Theory." *Qu'est-ce que la philosophie?* (Paris: Minuit, 1991), 144.

7. P. Benaceraff and H. Putnam, introduction to *Philosophy of Mathematics. Selected Readings,* 2nd ed. eds. P. Benaceraff and H. Putnam (New York: Cambridge University Press, 1983), 1.

8. See translation, p. 41.

9. Aristotle, *Metaphysics,* G, 1003a34.

10. Pierre Aubenque, *Le Problème de l'être chez Aristote* (Paris : PUF/Quadrige, 1962), 198.

11. Ibid., 190.

12. See translation, p. 159.

13. Alain Badiou and Peter Hallward, "Politics and Philosophy: An Interview with Alain Badiou," *Angelaki,* 3, no. 3 (1998): 130.

14. Alain Badiou, *Manifesto for Philosophy*, ed. and trans. Norman Madarasz (Albany: State University of New York Press, 1999), chap. 10.

15. Norman Madarasz, "A Potência para a simulação: Deleuze, Nietzsche e os desafios figurativos ao se repensar os modelos da filosofia concreta," in *Educação e Sociedade,* v. 26, n. 93, 2005, 1209–1216.

16. Gilles Deleuze and Felix Guattari, *Qu'est-ce que la philosophie?* (Paris: Minuit, 1991), 144.

17. See translation, p. 57.

18. See translation, p. 53.

19. See translation, p. 86.

20. He refers to this trajectory as the "Gamma Figure." Alain Badiou, "Conférence sur la soustraction," in *Conditions* (Paris: Editions du Seuil, 1992), 187.

21. Samuel Eilenberg and Saunders Mac Lane, "General Theory of Natural Equivalences," *Transactions of the American Mathematical Society,* 58 (1945): 231-94.

22. J. Bell, "Category Theory and the Foundations of Mathematics," *British Journal for the Philosophy of Science* 32 (1981): 356.

23. According to J. P. Marquis, "under the influence of William Lawvere the idea that category theory could be used as a foundation of mathematics came to be considered and developed seriously." Jean-Pierre Marquis, "Category Theory," *Stanford Dictionary of Philosophy (Spring 2004 Edition)*, Edward N. Zalta ed., URL = <http://plato.stanford.edu/archives/spr2004/entries/category-theory/>.

24. C. McLarty, "Uses and Abuses of the History of Topos Theory," *British Journal for the Philosophy of Science* 41 (1990): 354.

25. J. Bell, "Category Theory and the Foundations of Mathematics," *British Journal for the Philosophy of Science* 32 (1981): 357.

26. See translation, p. 145.

27. Norman Madarasz, "On Alain Badiou's Treatment of Category Theory in View of a Transitory Ontology," in *Alain Badiou: Philosophy and its Conditions,* ed. and with an intro. by Gabriel Riera (Albany: State University of New York Press, 2005), 23–43.

28. See translation, p. 151.

29. Alain Badiou, *Manifesto for Philosophy,* 65.

Prologue. God Is Dead

1. Ludwig Feuerbach, *The Essence of Christianity,* trans. George Eliot (New York: Prometheus Books, 1989).

2. S. Kierkegaard, *The Sickness Unto Death,* trans. A. Hannay (Toronto: Penguin, 1989), I. A, A.

3. Q. Meillassoux, *L'Inexistence divine* (forthcoming).

4. From the testimonial "Only a God can Save Us," *Der Spiegel* interview given in 1966, but only published posthumously. Richard Wolin ed., *The Heidegger Controversy,* (Cambridge: MIT Press, 1993).

5. Alvaro de Campos (a.k.a. Fernando Pessoa), *Selected Poems,* 2nd ed. trans. Jonathan Griffin (London: Penguin Classics, 1982).

6. Among the most important of contemporary Russian poets, Guenadi Nicolaievitch Aîgui (b. 1934) has written in both Russian and in his native Chuvash. (The Chuvash Autonomous Republic is part of the Russian Federation in Europe.) A victim of state censorship until Glasnost, his poetry is best known in Brazil, France, Italy, and Hungary. He remained unknown to Russian readers until the 1990s, despite prestigious awards received outside of the frontiers of the USSR. (My own translation is based on the French: in, *Hors commerce Aigui,* textes réunis et traduits du russe par André Markowicz (Paris: Le Nouveau Commerce, 1993)).

1. The Question of Being Today

1. M. Heidegger, "The Question concerning Technology," in *Basic Writings,* ed. and trans. David Farrell Krell (New York: Harper and Row, 1977), 311.

2. ———, *Nietzsche*, vol. 2: *The Eternal Recurrence of the Same*, trans. David Farrell Krell (San Francisco: HarperCollins, 1991.) (Translation altered).

3. G. W. Leibniz, "Letter to Arnaud" (April 30, 1687), *The Leibniz-Arnauld Correspondence*, ed. and trans. H.T. Mason (Manchester: Manchester University Press, 1967).

4. M. Heidegger, *Introduction to Metaphysics,* trans. Ralph Mannheim (Yale: Yale University Press, 1991), 45.

5. Stéphane Mallarmé, "Quant au livre: L'Action restreinte," in *Poésies. Anecdotes ou Poèmes, Pages diverses* (Paris: Le Livre de Poche, 1977), 210.

6. Lucretius, *On the Nature of Things.* I. 1000. (My translation is based on the French version used by Badiou.)

7. A. Badiou, *L'Etre et l'événement* (Paris: Editions du Seuil, 1986).

8. Plato, *Parmenides.*

9. Lucretius, *On the Nature of Things.* I: 425.

10. Plato, *The Republic,* VI, 511c.

11. Lucretius, *On the Nature of Things*, I: 910.

2. Mathematics is a Thought

1. Aristotle, *Metaphysics,* M. 2, 10. 1076b10.

2. Ibid., M. 3, 30; 1078a21.

3. Ibid., M. 3, 30; 1078b1.

4. Ibid., M. 3, 30; 1078a34.

5. Ibid., M. 3, 30; 1078a34–35.

6. Ibid., M. 2, 1077b1–2.

7. Nicolas Bourbaki was the name of the distinguished mathematics group centered in France which undertook to lay mathematics out clearly in terms of modern analysis based on Set Theory. The late Samuel Eilenberg (1913–1998), one of the creators of category theory, was part of the group.

8. Louis Althusser, *Philosophy and the Spontaneous Philosophy of the Scientists and Other Essays,* trans. Ben Brewster (New York: Verso, 1990).

9. See my introductory comments on adjusting translated citations to match the choices made by Badiou. Contrast the translation with "The same thing is there for thinking of and for being." Parmenides, *On Nature*, ed. Allan F. Randall from translations by David Gallop, Richard D. McKirahan, Jr., Jonathan Barnes, John Mansley Robinson, and others, http://home.ican.net/~arandall/Parmenides.

10. Isadore Ducasse (Comte de Lautréamont), *Songs of Maldoror*, II: 10, trans. Paul Knight (London: Penguin Classics, 1978). The extended verse reads: "Oh severe mathematics, I have not forgotten you since your wise lessons, sweeter than honey, filtered into my heart like a refreshing wave. . . . Without you, in my struggle against man, I would have perhaps been defeated." (Translation slightly altered.)

3. The Event as Trans-Being

1. French mathematician and philosopher, Albert Lautman (1908–1944) worked primarily in the 1930s. After the Nazi invasion of France, he joined the Résistance. In 1944, he was arrested and executed. His main body of work has been collected in, *Éssai sur l'unité mathématique* (Paris : Hachette, 1977), but has never been translated into English. Save for the most cosmopolitan historians of mathematics, Lautman remains pretty much unknown. Badiou draws a stirring portrayal of his life and work in the introduction to *Abrégé de métapolitique* (Paris: Éditions du Seuil, 1998). See also the special issue devoted to Lautman of: *Revue d'histoire des sciences* (Paris : Presses Universitaires de France, 1987) 40.

2. Alain Badiou, *Le Nombre et les nombres* (Paris : Éditions du Seuil, the "Des travaux" series, 1990), chap. 12.

3. Lucretius, *On Nature*, II: 496.

4. Deleuze's Vitalist Ontology

1. A. Badiou, *Deleuze: The Clamor of Being*, trans. Louise Burchill (Minneapolis: University of Minnesota Press, 1999). Regarding the "reproaches," see the essays by Eric Alliez, Arnaud Villani and José Gil, in "Dossier Badiou/Deleuze," in *Futur Antérieur*, 43: April 1998.

2. G. Deleuze and F. Guattari, *A Thousand Plateaus*, trans. and with a foreword by Brian Massumi (Minneapolis: University of Minnesota Press, 1987), 344.

3. G. Deleuze, *Difference and Repetition,* trans. Paul Patton (New York: Columbia University Press, 1994), 37.

4. See A. Badiou, *Deleuze: The Clamor of Being,* op. cit., chap. 22: "The 'Purified Automaton.' "

5. G. Deleuze, *The Movement-Image,* trans. Hugh Tomlinson and Barbara Habberjam (Minneapolis: University of Minnesota Press, 1986), 10. (Translation altered).

6. G. Deleuze and F. Guattari, *What is Philosophy?* trans. Hugh Tomlinson and Graham Burchell (New York: Columbia University Press, 1994), 60.

5. Spinoza's Closed Ontology

1. B. Spinoza, *Ethics*, p 34.1. My translations are based on G. H. R. Parkinson (New York: Everyman Classics, 1989).

2. Spinoza, *Ethics*, p 18.1.

3. Ibid., p 7.2.nn.

4. Bernard Pautrat's French translation of Spinoza's *Ethics* was published by les Éditions du Seuil, Paris, 1988. An equivalent to the verb form of *"intellect"* does not exist per se in English. R. H. M. Elwes uses it as a noun in his translation of the letter to De Vries, cited below, despite preferring "understand" when a verb in his translation of *Ethics,* p 40.5.nn. (See notes 6 and 21.)

5. Spinoza, *Ethics,* d4.1

6. 9(27). Spinoza to De Vries, February 1663. (My translation is based on *EPISTOLÆ doctorum quorundam virorum ad B. d. S. et auctoris RESPONSIONES; ad aliorum ejus operum elucidationem non parum facientes,* (trans. R. H. M. Elwes, 1677). Numbers in accordance with Gebhardt (86 Letters, including 48bis and 67bis), in parentheses-number in Opera Posthuma.

7. Spinoza, *Ethics,* p 31.1.

8. Spinoza, *Ethics,* after p 21.1.

9. 64(66). Spinoza to Schuller, July 1675.

10. 32(15) Spinoza to Oldenburg, November 1665.

11. Spinoza, *Ethics,* p 16.1.

12. Ibid., p 13.2.nn.

13. Ibid., *Ethics*, p 2.3.

14. Ibid., *Ethics*, after p 21.2.

15. Ibid., *Ethics*, a 6.1.

16. Ibid., *Ethics*, d 4.2.

17. Ibid., *Ethics*, p. 21.2.nn.

18. Ibid., *Ethics*, p 7.2.

19. Ibid., *Ethics*, p 22.1.

20. Ibid., *Ethics*, p 11.2.c.

21. Ibid., *Ethics*, p 40.5.nn.

22. Ibid., *Ethics*, p 24.2.

23. Ibid., *Ethics*, p 19.2.

24. Ibid., *Ethics*, p 43.2.

25. Ibid., *Ethics*, p 43.2.

26. Spinoza, *Ethics*, lemma 1 after p 40.2.

27. Ibid., *Ethics*, after p 44.2.cc.

6. Platonism and Mathematical Ontology

1. P. Benacerraf and H. Putnam eds., *Philosophy of Mathematics. Selected Readings,* 2nd ed. (New York: Cambridge University Press, 1983), 18.

2. Parmenides, 3. See footnote 2.9 above.

3. A. A. Fraenkel and Yehoshua Bar-Hillel, *Foundations of Set Theory* (Amsterdam: North-Holland Publishing Company, 1958).

4. K. Gödel, "What is Cantor's continuum problem?," in *Frege and Gödel. Two Fundamental Texts in Mathematical Logic,* ed. Jean van Heijenoort, (Cambridge, MA: Harvard University Press, 1970), 484–85.

5. L. Wittgenstein, *Tractatus Logico-Philosophicus,* trans. C. K. Ogden (London: Routledge and Kegan Paul, 1922), 6:211.

6. See A. Badiou, *L'Être et l'événement* (Paris: Éditions du Seuil, 1988), "Méditations" 31–34.

7. The Aristotelian Orientation and Logic

1. G. W. Leibniz, *New Essays on Human Understanding,* eds. P. Remnant, J. Bennett, K. Ameriks, and Desmond M. Clarke (London: Cambridge University Press, 1996), 157.

2. Badiou has announced that the theory of categories and *toposes* will be examined in detail in his forthcoming book, *Logiques des mondes.*

8. Logic, Philosophy, "Linguistic Turn"

1. M. Heidegger, "What are Poets for?," *Poetry, Language, Thought,* trans. Albert Hofstadter (New York: Harper and Row, 1975).

2. Ludwig Wittgenstein, *Culture and Value,* ed. G. H. Von Wright (in collaboration with Heikki Nyman), trans. Peter Winch (Oxford: Basil Blackwell, 1980), 24e.

3. Alain Badiou, *Manifesto for Philosophy,* chap. 5.

4. F. Nietzsche, *Thus Spoke Zarathoustra,* trans. R.J. Hollingdale (London: Penguin, 1969), 153–154.

5. S. Eilenberg and S. Mac Lane, "General Theory of Natural Equivalences," *Transactions of the American Mathematical Society* 58 (1945): 231–94.

6. Alexander Grothendieck, and J. L. Verdier, "Topos," in *Théorie des Topos et Cohomologie Etale des Schémas SGA 4,* ed. M. Artin, A. Grothendieck, and J. L. Verdier 269 (Springer Lecture Notes in Mathematics, Springer Verlag, 1972), 299–519.

7. P. Freyd, "The theory of functors and models," in *The Theory of Models,* eds. J. W. Addison, L. Henkin, and A. Tarski (North-Holland Pub., 1965), 107–20; F. W. Lawvere, "The Category of Categories as a Foundation for Mathematics," in *Proceedings of the La Jolla Conference in Categorical Algebra* (Springer-Verlag, 1966), 1-20.

8. Jean-Toussaint Dessanti, "A Propos de l'ontologie intrinsèque d'Alain Badiou," *Les Temps modernes* 526 (mai 1990): 61–71.

9. First Remarks on the Concept of *Topos*

There are no notes in this chapter.

10. First Provisional Theses on Logic

1. Leibniz, *The Monadology,* ed. and trans. Robert Latta, sec. 47 (London: 1898), http://www.rbjones.com/rbjpub/philos/classics/leibniz/monad.htm.

2. Stéphane Mallarmé, "Igitur," in *Selected Poetry and Prose,* ed. and trans. Mary Ann Caws (New York: New Directions, 1982).

3. S. Mallarmé, "Ses Purs ongles très hauts dédiant leur onyx" [Her Pure Fingernails"] (1887), http://poesie.webnet.fr/poemes/France/mallarme/3.html. (My translation with the "ix" and "or" in italics.)

11. The Being of Number

1. Surreal numbers are "the most natural collection of numbers which includes both the real numbers and the infinite ordinal numbers of Georg Cantor. They were invented by John H. Conway in 1969. Every real number is surrounded by surreals, which are closer to it than any real number. The surreal numbers are written using the notation {a|b}, where {|}= 0, {0|}= 1 is the simplest number greater than 0, {1|}= 2 is the simplest number greater than 1, and so forth. Similarly, {|0}= −1 is the simplest number less than 0, and so forth. However, 2 can also be represented by {1|3}, {3/2|4}, {1|w}, and so forth." Eric W. Weisstein, "Surreal Number," from *MathWorld*—A Wolfram Web Resource, http://mathworld.wolfram.com/SurrealNumber.html.

2. The late Gilles Châtelet notably published *Les Enjeux du mobile. Mathématique, physique, philosophie* (Paris : Seuil, the "Les Travaux" series, 1999.) Badiou published an enthusiastic review of this work in "Les Gestes de la pensée," *Les Temps Modernes* 586 (1996): 196–204.

3. G. W. Leibniz, *Monadology,* trans. Robert Latta, sec. 67.

12. Kant's Subtractive Ontology

1. Immanuel Kant, *Critic of Pure Reason,* trans. Norman Kemp Smith (London: Macmillan Press, 1929), B130–131.

2. Ibid., B134.

3. Ibid., B135.

4. Ibid., B143.

5. Ibid., A111.

6. Ibid., B156.

7. Kant, *Critique of Pure Reason*, B156 ff.

8. Ibid., B156 ff.

9. Ibid., B156 ff.

10. Ibid., B156 ff.

11. Heidegger, *Kant and the Problem of Metaphysics,* trans. James S. Churchill (Bloomington, IN: Indiana: Indiana University Press, 1962).

12. Ibid.

13. Kant, *Critique of Pure Reason*, B138.

13. Group, Category, Subject

1. Jean-Toussaint Desanti, "A Propos de l'ontologie intrinsèque d'Alain Badiou," *Les Temps modernes* 526 (mai 1990): 61–71.

14. Being and Appearing

1. Kant, *Critique of Pure Reason,* trans. Norman Kemp Smith (London: Macmillan Press, 1929), B: ix.

2. Claude Imbert, 1978. "Théorie de la représentation et doctrine logique dans le stoïcisme ancien," in *Les Stoiciens et leurs logiques,* ed. J. Brunschwig (Paris : Vrin, 1978), 223–49.

3. In the French original, Badiou leans on a recently acclaimed French translation of Aristotle's *Metaphysics* by Barbara Cassin and Michel Narcy. I have eliminated his direct reference to the translation in the text, though I have calibrated the English translation of the *Metaphysics* to the results achieved by Cassin and Narcy in French. M. Narcy (en collaboration avec Barbara Cassin), *La décision du sens. Le livre* Gamma *de la Métaphysique d'Aristote*, introduction, texte, traduction et commentaire. «Histoire des doctrines de l'Antiquité classique» 13 (Paris : Librairie philosophique J. Vrin, 1989). Aristotle, *Metaphysics,* Γ, 1003a28.

4. Ibid., Γ 1005b20

5. Ibid., Γ, 1011b24.

6. The term "*pros hen*" appears in *Metaphysics, Γ,* 4. As with most key Greek concepts used by Badiou, we deal not only with the cultural difference between French translations of ancient Greek, but also with his own account of Aristotle's speculative science of beings *qua* beings. For instance, the *Stanford Dictionary of Philosophy* offers the translation of *pros hen* as "in relation to one," an idea quite in line with the analytic tradition. As we shall see, it only partially matches Badiou's reading.

7. In reference to French Revolution leader and infamous head of the Comité du Salut Public, Maximilien Robespierre's speech at the Jacobin Club on January 8, 1794, in which he declaimed against "those who of an ardent genius and exaggerated character propose *ultra*-revolutionary measures, and those who of a milder manner and more moderate propose *citra*-revolutionary means. They fight against each other, but whether either one or the other ends up victorious, this barely interests them. As either system leads to losing the Republic, they generally achieve a successful result." (My translation).

8. I. Kant, *Critique of Pure Reason,* trans. Norman Kemp Smith (London: Macmillan Press, 1929), A426–7/B454–455.

CONTRIBUTORS

Alain Badiou was professor of philosophy at Université Vincennes (currently: Université Paris 8—Vincennes à Saint-Denis) from 1969 to 1999. Since 1999, he has held the Chair of Philosophy at the École normale supérieure in Paris. Professor Badiou has published over fifteen books, including *L'Etre et l'événement,* and *Deleuze: The Clamor of Being.* His work has been translated into several languages, and he has lectured in various countries. His political group, *L'Organisation politique,* regularly publishes a pamphlet on labor affairs in France. With Barbara Cassin, he is editor of the "L'ordre philosophique" series at Les Éditions du Seuil. Professor Badiou is also an accomplished playwright and novelist. He is currently completing work on a sequel to *L'Être et l'événement,* called *Logique des mondes.*

Norman Madarasz (Docteur de philosophie) is Visiting Associate Professor of Philosophy at Universidade Gama Filho (*bolsista da CAPES*), in Rio de Janeiro, Brazil. He has edited and translated Alain Badiou's *Manifesto for Philosophy,* (Albany: State University of New York Press, 1999). Professor Madarasz has published numerous essays, articles, and reviews, and is currently completing a book on faith in the context of atheism. He also writes on political economy, international relations and the arts for *CounterPunch, Islam on Line, Brazzil,* and other journals.

INDEX

Abraham, 23
Act, 123
Aîgui, Guennadi Nicolaievitch, 31, 173n6
Althusser, Louis, 3, 51
Anderson, Perry, 2
Anglophone world, 17, 101
Appearing (*l'apparaître*), 152–168
    passim
Aron, Raymond, 2
Aristotle, xi, 13, 45, 47, 60, 98,
    101–106, 110, 157, 164
  Aristotelian orientation, 49, 97
  conception of God, 25
  logic, 154–155
  mathematika, 45, 158–159
  Metaphysics, 45, 155, 157
  wish for an ontologically prescribed
    logic, 167–168
Atheism
  contemporary, definition, 29
Aubenque, Pierre, 5
Axiom(s), 42, 93
  axiomatic decision, 119–120
  axiomatic thought, 38–40
  choice, of, 52, 96

Bar-Hillel, Yeroshua, 91–92
Beckett, Samuel, ix, 3, 4, 109
Being (*Être, Sein*), 33, 42, 52, 97–98,
  153–168 passim
  collapse of consistency, 168
  distinct from appearing, xii
  enframing by the One, 34
  its inexistence as whole, 160

oblivion of, 28
physis, 101
qua Being, 7, 43, 59–61, 69, 87,
  107, 131, 153, 159, 163–167
  related to experience, 57
  thought of, the, xi
  See also Dasein, Heidegger
Beings (*étant(s), seinend(s)*), 161–162
  See also Dasein, existents
Bell, J., 13
Benaceraff, Paul, 5, 89
Bergson, Henri, 115, 117, 122
Body, 27
Boole, George, 107, 164
Bourbaki, Nicolas, 49, 174n7
Bourdieu, Pierre, 2
Brunschvicg, Leon, 160

Cahiers pour l'analyse, 4, 16
Cambodia, 2
Cantor, Georg, 5, 22, 30, 40, 96, 105,
  123, 131, 165
  Continuum Hypothesis, 10, 76, 99
  ordinals, 126
Carnap, Rudolf, 154
Cassin, Barbara, 180n3
Category theory, x, 12, 105, 113,
  165–166
  functors, 12, 13
  morphism(s), arrows, 145–148, 167
  One, identity, isomorphism, 146–148
  sheaves, 14
  set, 149
  topos, 110–111, 115–117, 166–167

Cercle d'épistémologie de l'E.N.S., 4, 16
Certeau, Michel de, 23
Châtelet, Gilles, 128, 179n2
Christ
  resurrection of, 21
Cohen, Paul, 99, 105
Conway, John Horton, 60, 127

Dasein, 121
  Being-for-death, 6, 30
  pros hen, 6, 157–168 passim, 181n6
Decision (in thought), 104
Dedekind, Richard, 131
Definition, 37
Deleuze, Gilles, x–xi, 3, 63–76 passim, 98, 115, 117
  conception of multiplicity, 16
  disjunctive synthesis, 66
  event, 61, 70
  feminitude, 63
  fold, 76
  inversion of Platonism, 8
  life as Being's main name, 62, 64, 68, 73
  logic, 122
  ontology, 16
  rejection of ontological mathematicity, 62
  rhizomatics, 50
  set, 69
  simulation, 17
  univocality of Being, 16
  virtuality, 63
Democritus, 35
Desanti, Jean-Toussaint, 4, 110, 145
Descartes, René, 16, 22, 25, 105
  dualism: mind-body theory, 16
  mathematics, 16
  subject, 16
Difference, 116
Dolto, Françoise, 23

Eilenberg, Samuel, 12, 110, 166
Element
  to be an, 7
Essentialism, 15, 17

Être et l'événement, L,' (Badiou), ix, x, 3–4, 7, 10–11, 35, 54, 56, 59, 152, 160–161, 183
Euclid, 50, 125
Eudoxus, of Cnidus, 52
Event, 60, 168
Excluded middle, 116
Existence, 9, 45–57 passim, 103
Existent, 6, 161–162
  See also Being, beings

Faith, 25
Feuerbach, Ludwig, 22
Finitude, 29, 124
Foucault, Michel, 1
France/French
  hermeneutics, 3
  in the 1960s, 2
  in the 1970s, 2
  structuralism, 3, 14
  sans papiers, les, x–xi
  thought as distinct from Anglo-American, 19
Fraenkel, Abraham, 91
Fraud, 19
  See also plagiarism
Frege, Gottlob, 7, 51, 107, 112, 130, 164
Freud, Sigmund, 150
Freyd, P., 110
Fundamentalism: forms of, 24, 26
Furet, François, 2

Galilei, Galileo, 22
Gang of Four, the, 2
Generic, 86
  See also manifold, truth
God(s), 10
  Christian God: death of, xi, 21–32; living God, 22–32; "God is Dead," xi, 22
  of Metaphysics, the First-Principle God, 25–31
  of poets, 28–32 passim
  of religion, 24–32 passim
  disappearance without return, 30

Name-of-the-Father, 22
See also Aristotle, Leibniz, Spinoza
Godard, Jean-Luc, 2
Gödel, Kurt, 92–96 passim, 107, 164
  Completeness Theorem, 11
  doctrine of constructible sets, 56
  mathematical intuition, 93, 95, 99
Grothendieck, Alexandre, 14, 110
Group(s), 143–149 passim
  definition in categorial terms, 148
  structure, 143
Guattari, Félix, 171n6

Hamlet (Shakespeare), 24
Hegel, G.W.F., 17, 60, 110, 153
Heidegger, Martin, xi, 1, 5, 7, 26–28,
    33, 108, 140
  Earth, what the poem names, 109
  event (*Ereignis*), 61
  linkage, 135–136
  logic, 155–156
  mathematics, 108
  on Kant, 132, 135, 138
Hilbert, David, 42, 107
Hölderlin, Friedrich, 28, 109
Humanism
  Post-Cartesian, 28
Hume, David, 136, 140
Husserl, Edmund, 42
Huxley, T. S., 16

Identity
  being identical to oneself, 141
  as opposed to coupling, 150
Imbert, Claude, 155
Infinite, 30, 104
Isaac, 23

Jacob, 23
Jospin, Lionel, 29

Kant, Immanuel, xi, 130–141 passim,
    163
  combination (*Verbindung-liaison*), 133
  counting, as correlation of two
    voids, 137, 139
  logic, 154–156

object of experience, 135–136
object, transcendental, 137
One, law of the, 134
original apperception, 134–138
    passim
  subject, transcendental, 137
  supersensible, 140
  unity (*Einheit*), 133, 139
Kierkegaard, Soren, 24
  antiphilosopher, the, 25
Kleene, 164
Kripke, Saul, 16
Kronecker, Leopold, 130

Lacan, Jacques, ix, 4, 40, 47
  coupling of the Real and the
    Imaginary, 151
  event, 61
  non-existence of a metalanguage,
    120
  religious belief, 23
  symbolic, the, 151
Lacoue-Labarthe, Philippe, 30
Lautman, Albert, 60, 175n1
Lautréamont, Isadore Ducasse Comte
    de, 57, 71
Lawvere, F.W., 110
Letter, the, 151
  See also psychoanalysis
Lenin, Vladimir Ilyich, ix
Leibniz, G.W., 34, 101–105, 115,
    117, 164
  his God, 119–120, 166
  Monadology, 130
Lévi-Strauss, Claude, 143
Linguistic Turn (of Western philoso-
    phy), 66, 96, 107, 159, 165
  break with, 111, 120
  dominion of language, 119
Limit, 97
  immanence of, 36
Logic, 106, 120, 122, 153–168 passim
  formalized, xii
  knotted to mathematics, 158, undoing
    the knot, 158
  mathematization, 111, 155
  possible universes, 164

"Logistics," 160
Lucretius, 5, 35–36, 40, 62, 74, 76
    Clinamen, 73

Mac Lane, Saunders, 12, 110, 166
McLarty, Colin, 14
Mallarmé, Stéphane, ix, 4, 35, 47, 71,
    109, 122–124, 168
    *Dice Thrown Will Never Annul
        Chance,* 123
    Igitur, 123
Man
    name of, 28
Manifold(s), 31, 38, 109, 139, 160–
    168 passim
    -being, 30, 73, 95, 162
    generic, xii, 11, 86
    See also multiplicity, truth
Mao
    See Zedong, Mao
Marquis, Jean-Pierre, 172n23
Mathematics, 43, 56, 103, 105, 113,
    154, 165
    condition of philosophy, 54
    crisis in, 54, 59
    formal science, 104
    history of, 159
    relation to philosophy, xii, in
        "de-relation" to logic, 113
    thought on Being *qua* Being, 165
    = ontology, identity thesis of, 54, 59
    See also manifold, matheme,
        multiplicity, number theory,
        ontology, "philosophy of
        mathematics"
Matheme, 86, 150
Meillassoux, Quentin, 24
Melville, Herman, 66
Multiplicity
    manifold, multiple, 7, 36–41
        passim, 109
multiple-without-One, 35, 41
    inconsistent, without a predicate,
        39, 40

Nancy, Jean-Luc, 23
Narcy, Michel, 180n3

Neumann, John von, 130
Nietzsche, Friedrich, x, 26–27, 65–70
    passim, 122
    anti-Platonism, 101
Norm, 38
Number(s)
    being of, 125
    cardinals, large cardinals, 97
    definition of, 126–131
    gesture in Being, 131
    irrational, 50
    macro-group, 129
    ontology of, 124
    real, 165
    surreal, 97, 127

One (the), 6, 30, 33, 35–36, 150, 157
    One and Multiple/Many, 65, 73
    Parmenidean tradition, 117
Ontology, 38–40
    mathematics, 159
    science of Being *qua* Being, 166–167
    subtractive, 139
    transitory, v, xi, 5
    See also mathematics
Onto-logy, 99, 105, 115, 119, 145
    as logic of Being, 18
Orientation in thought, 11
    three types, 55
    definition, 53–54
    See also truth procedures
Ortega y Gasset, José, 1
Other (the), 36, 98, 151
    See also Plato: *Sophist*

Parmenides, 52, 90, 175n9
Pascal, Blaise, 23, 25, 60
Paul, Saint, x, 23
Pautrat, Bernard, 75, 176n4
Peano, 130
Pessoa
    heteronyms: Campos, Alvaro de,
        30; Caiero, Alberto, 30
Phallus, 27
Philosophy, 60–61
    dawn of, 150
    ordinary language, 159

"Philosophy of mathematics," 106
Plagiarism, 19
  See also fraud
Plato, xi, 7, 10, 13, 15, 17, 33, 45,
    64, 87, 91, 105, 115, 158
  Cratylus, 159
  Idea, 94, 102
  Meno, 89
  Parmenides, 36, 42, 60, 98
  Philebus, 42
  "Platonic gesture," 7, 33
Platonism: spontaneous philosophy of
    mathematicians, 48–49, 89;
    definition, 91–96 passim, 102
  Republic, 38
  slave owners' ideologue, 101
  Sophist, 42, 98
  five supreme genera, 149
Poem
  "Age of Poets," 109, 122
  its destiny, 29
Poincaré, Henri, 160
Possible, 120
Postmodernism, 19
Psychoanalysis
  cure (of the subject), 151
  letter, the, 151
  See also Lacan, Jacques
Putnam, Hilary, 5, 89
Pythagoreans, 50, 52

Ramsey Cardinal, 96
Real, the, 30, 45, 56, 102, 120, 122,
    159
Relation, 5, 65, 70, 139, 152, 163
Renaissance, 22
Renan, Ernst, 22
Revolution, 2
Rimbaud, Arthur, ix, 23, 109
Robespierre, Maximilien, 159, 181n7
Romanticism, German, 10
Rousseau, Jean-Jacques, 60
Rowbottom, F., 96
Russell, Bertrand, 107
  logicism, 108
  paradox of, 7, 51, 52

Saint-Just, ix
Same, the, 149, 151
Sartre, Jean-Paul, 2
Sense, 29
Set theory, x, 5, 10, 12, 42, 56, 97,
    105, 112, 120, 127, 160
  Zermelo-Fraenkel with axiom of
    choice (ZFC), 7
  Universe, single, 15
  Void, empty set, 98, 117
  See also Cantor, manifold(s)
Situation, 161, 164
Socrates
  See Plato
Spinoza (Baruch), xi, 10, 18, 60–87
    passim
  "the Christ of philosophy," 71
  causality: the first fundamental
    relation, 73–82 passim
  common objects, 84
  coupling: the second fundamental
    relation, 75, 79–80
  finite intellect, 83, 85
  freedom, 86
  God, 73–85 passim
  inclusion: the third fundamental
    relation, 75, 81
  infinite, 78
  intellectus (*intellecter:* to intellect),
    75–77
  more geometrico (geometrical
    manner), 74, 86
  true knowledge, 84
  See also "there is"
State, 31
Stoics
  logic, 155
Subject, x, 16, 21, 89, 90, 123, 141,
    162
  as plait, 150
  group as mathem for a thought on,
    150, as infinite, 151
  its logic, 141
  the obscure, 27
Substance (infinite), 71, 74
  See also Spinoza
Symmetry, 150

System, systematicity, 17
    notion of, 12

Tarski, Alfred, 164
Temps modernes, les, 4
"There is/are," the (*le "il y a";"es gibt"*), 40, 73–76 passim, 86–87, 135, 139, 163
Timor, East, 2
Transcendence, 7
Translating, 1, 18
Truth, 11, 93
    category of, 122
    no truth, 61
    See also manifold(s), orientations
Truths, xii, 61
    See also manifold(s)

Undecidable, 92, 94
United States of America, the
    Air Force: bombings, 2

Universe, 13
    See also possible worlds

Void, 6, 117, 168

Whitehead, Alfred North, 66
Wittgenstein, Ludwig, x, 42, 45, 94, 115, 130
    mystical element, 61
    non existence of a metalanguage, 121
    private-language argument, 16
    world and language, 108–110
World(s)
    Possible, 13
    Single, 15

Zarathustra, 69, 110
Zedong, Mao, ix
Zermelo, Ernst, 53, 92
    See also Set Theory